GENERATIVE AI TRANSFORMATION BLUEPRINT

BYTE-SIZED LEARNING SERIES
BOOK 3

I. ALMEIDA

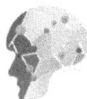

NOW
NEXT
LATER

We are the most trusted and effective learning platform dedicated to empowering leaders with the knowledge and skills needed to harness the power of AI safely and ethically. Join now to enjoy free lessons and webinars, and check out other books in the series.

CONTENTS

1

INTRODUCTION

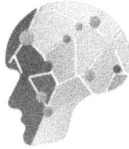

The Advent of Generative AI

Artificial intelligence has advanced tremendously in the last decade from narrow domain-specific capabilities to more expansive, multi-functional systems that can synthesize novel artifacts like text, images and video with increasing sophistication. This new era heralded the dawn of generative AI.

Unlike previous reactive AI systems designed for tasks like visual recognition or predictive analytics, generative models create completely new data patterned on training datasets. Groundbreaking examples include systems like DALL-E which generate striking images simply from text descriptions or the GPT series capable of crafting synthetic but cogent essays on arbitrary topics.

Beyond their almost magical creativity, these emergent capabilities presage significant shifts across industries. As generative AI proliferates, competitive advantage will center on effectively leveraging its applications. However, doing so calls for prudent strategy stemming from insightful examination of suitable use cases, thoughtful approaches for scalable implementation and anticipation of risks from model opacity or data privacy threats.

This transformation demands demystification for leaders navigating uncharted waters filled with hype, false starts and uncertainty about best practices.

Demystifying Generative AI Transformation

This guide provides senior decision-makers with a clear, accessible roadmap for harnessing the power of generative AI, enhancing innovation, and boosting business outcomes. Drawing on insights from leading consultancies and input from both established and rising leaders in the AI field, it presents a validated strategic approach. This blueprint not only outlines best practices but also showcases pioneering use cases, integrating them into a cohesive framework for practical implementation.

Spanning across areas from strategic alignment and talent development to ethical governance and sustaining competitive edge amid relentless underlying progress, it delivers clarity for charting an optimal generative AI roadmap.

Who this Book is For

The core audience comprises senior executives like CEOs, strategic planners, technology heads, product leaders or functional unit heads keen on harnessing generative AI for a competitive edge but needing authoritative counsel consolidating recent lessons into a crisp actionable package to aid planning.

How this Book is Structured

The chapters provide end-to-end coverage beginning with foundational concepts, leading into implementation modules and culminating in sustenance best practices:

Chapters 2-4 establish understanding, discovery mindsets and strategic alignment principles constituting base bricks for subsequent generative AI build phases.

Chapters 5-8 guide technology, infrastructure and capability upgrades for pilot testing with protocols for systematizing scaling.

Chapters 9-11 cement responsiveness and innovation elements needed for maximizing generative AI reliability and longevity despite external flux.

Chapter 12 offers concluding thoughts on the road ahead.

All chapters include sample scenarios and helpful frameworks and research, convenient for reference during planning.

With expansive technological disruption on the horizon, this handbook delivers a visionary blueprint for leadership teams to harness generative AI as a catalyst for unprecedented progress. Let us turn the page and begin this transformative journey.

UNDERSTANDING AND AWARENESS

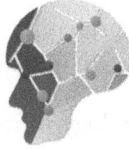

Generative AI represents a seismic shift in artificial intelligence capabilities. Systems that were previously focused narrowly on specific tasks can now perform a much wider range of cognitive functions in an increasingly human-like manner. This poses both tremendous opportunities and complex challenges for organizations seeking to leverage these rapidly evolving technologies.

As generative AI permeates across industries, business leaders require a foundational understanding of its potentials and limitations to chart an effective strategic course. This chapter aims to demystify key aspects of generative AI and establish best practices for continuous learning and benchmarking. With comprehensive understanding and awareness, organizations can make informed decisions to harness generative AI as a transformative driver of innovation and growth.

Demystifying Generative AI

WHAT IS GENERATIVE AI?

Generative AI refers to a class of artificial intelligence algorithms capable of producing novel, high-quality artifacts with little or no human guidance. The term encompasses a range of techniques including generative adversarial networks, diffusion models, reinforcement learning, and transformer architectures.

Unlike traditional AI systems designed for narrow tasks like classification and prediction, generative models can synthesize various kinds of data such as text, images, video, and audio that capture intricate statistical patterns from their training data. Prominent examples of generative AI today include systems like DALL-E for image generation and the GPT series for natural language processing.

CAPABILITIES AND APPLICATIONS

The open-ended nature of generative models unlocks promising new capabilities across diverse domains:

- Natural language processing: Automated writing, conversational systems, language translation

- Computer vision: Image and video generation, editing media

- Drug discovery: Identifying potential therapeutic molecules

- Design: Generating logos, websites, industrial design blueprints

- Personalization: Customized marketing content, personalized recommendations

Leading technology research firm Gartner predicts that by 2025, 70% of enterprises will use some form of generative AI to augment business operations, a significant leap from less than 5% in 2022.

STRENGTHS AND PROMISE

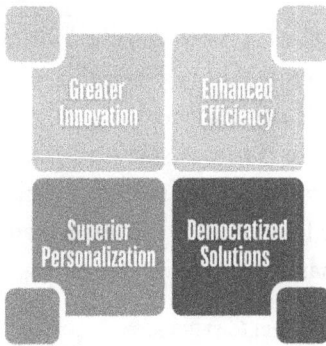

Generative AI offers organizations substantial benefits, including:

- Greater Innovation—By automating complex creative tasks, generative AI exponentially expands an organization's capability to experiment, ideate and produce novel solutions. For instance, generative design tools can rapidly output thousands of options for a new product component.

- Enhanced Efficiency—Routine tasks like writing reports or designing documents can be automated to enable employees to focus their efforts on higher-value work. Generative AI can also enhance decision-making by rapidly synthesizing and analyzing vast amounts of data.

- Superior Personalization—Fine-tuned models can capture granular insights about customer preferences and behaviors to create tailored solutions. From personalized medicine to customized marketing, the applications span across sectors.

- Democratized Solutions—Pre-trained generative models encapsulate capabilities that can be readily tapped by users without specialized machine learning expertise. This democratization effect makes AI accessible beyond data scientists.

LIMITATIONS AND CHALLENGES

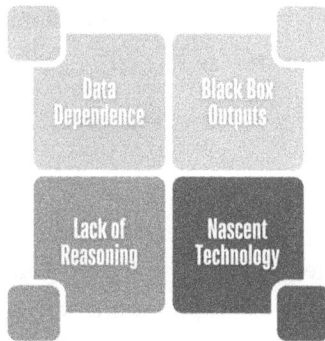

While promising, generative AI continues to have key limitations organizations should recognize:

- Data Dependence: Performance hinges completely on the system's training data quality and distribution. Skewed or low-quality data readily leads to biased and unreliable outputs.

- Black Box Outputs: The stochastic nature of generative algorithms means results can be unpredictable. Post-hoc analysis is needed to detect potential inaccuracies or bias.

- Lack of Reasoning: Current techniques excel mainly in pattern recognition abilities. Capabilities requiring complex reasoning, contextual understanding or manipulation of abstract concepts remain limited.

- Nascent Technology: Most generative AI today necessitates careful human oversight and judgment. Full autonomy across complex tasks is still a distant prospect in emerging fields like drug design or engineering.

LOOKING AHEAD

The generative AI field continues to witness explosive progress. With sustained investments and advances in areas ranging from energy-efficient computing hardware to next-generation algorithms, expectations are high for revolutionary developments in coming years.

Gartner projects[1] that by 2025, generative AI will account for 10% of all data produced, up from less than 1% in 2022—on par with capabilities like IoT devices and mobile apps. As the technology proliferates across industries, future shifts may include AI systems attaining creative proficiency in specialized domains, enhancements in contextual and reasoning abilities, as well as increased trust and adoption of autonomous generative applications.

Research and Benchmarking

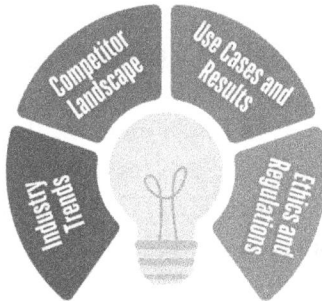

To fully harness generative AI's disruptive potential while managing risks, organizations need an in-depth understanding of trends, use cases and best practices. This necessitates dedicated efforts for ongoing research and benchmarking. Critical focus areas include:

- Industry Trends: Continuously tracking the pulse of the generative AI landscape is crucial for updated technology awareness and early identification of emerging opportunities. Research should cover developments across core techniques, new model architectures, advances in hardware optimizations as well as real-world adoption patterns across industry verticals.

- Competitor Landscape: Analyzing strategic moves and implementations by rivals in the competitive space spotlights innovative applications and gives intelligence on where to focus investments for maximum impact.

- Use Cases and Results: Studying empirical examples of generative AI deployments across diverse domains provides tangible insights on practical implementation challenges, guidelines for customization and key factors driving ROI.

- Ethics and Regulations: Monitoring the regulatory policy landscape and debates on ethical considerations provides foresight into potential legal or reputational risks associated with generative AI. It also gives guidance on governance best practices.

Strategic Focus Areas

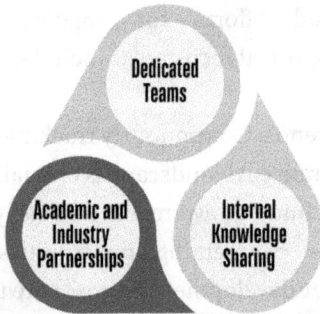

To build effective research and benchmarking capabilities, organizations should prioritize:

- Dedicated Teams: Allocate resources expressly focused on generative AI research and analysis. Given the rapid pace of change, this area merits specialized attention.

- Academic and Industry Partnerships: Collaborating with external ecosystem partners through sponsored university research, incubator projects, hackathons and more brings access to breakthroughs at the cutting edge.

- Internal Knowledge Sharing: Ensure insights from research get disseminated company-wide through vehicles like AI Expert Talks, demo days, wikis and engineering blog posts.

THE WAY FORWARD

With comprehensive understanding grounded in continuous learning and benchmarking, organizations can tap generative AI as a force multiplier and source of competitive advantage. However, as recent controversies highlight, merely chasing the state-of-the-art blindly without accounting for reliability gaps or ethical dilemmas also poses serious perils.

Constructive progress necessitates building institutional knowledge coupled with responsible governance and oversight mechanisms. The strategies in this chapter reinforce that foundation—serving as indispensable guideposts on the path to harnessing generative AI effectively.

Sample Cases in Action

To bring the concepts of this chapter to life and demonstrate their practical application in various industries, I present a series of illustrative use cases. These cases, drawn from a range of sectors and organizational contexts, are designed to help readers envision real-world scenarios. They offer insights into potential opportunities and risks, suggest strategies for mitigation, and guide on evaluating outcomes. Through these examples, readers can better understand how to apply the principles discussed in this chapter to their own unique situations and challenges.

SAMPLE CASE 1: DEMYSTIFYING AI FOR LEADERSHIP TEAMS

Industry: Financial Services

Company Size: Large Enterprise

Business Scenario: A leading investment bank sought to educate its senior leadership on AI capabilities to aid strategy planning. However, most lacked technical backgrounds, necessitating demystification.

Solution: The bank designed a customized 2-day offsite workshop. Sessions covered introductory concepts using case studies, frank discussions on limitations and an outlook on long-term potentials.

Outcomes and Impact: Post-workshop surveys indicated a sharp rise in AI comprehension among leaders, laying the foundation for informed strategic planning. 80% of attendees rated the workshop as very effective.

Implementation Challenges: Securing full executive team participation due to scheduling complexity. Addressed through calendar prioritization.

SAMPLE CASE 2: CONTINUOUS AI TREND TRACKING

Industry: Healthcare

Company Size: Mid-sized Organization

Business Scenario: A healthcare technology company sought to continuously monitor AI developments to spot promising innovations applicable to its diagnostic solutions suite.

Solution: The company established an interdisciplinary Research & Benchmarking team with data scientists, clinicians and strategy leads. This team produces periodic landscape reports, hosts tech talks and scans use cases.

Outcomes and Impact: The initiative has sparked several experiments with natural language processing and computer vision techniques which are garnering strong industry interest.

Implementation Challenges: Researchers occasionally pursuing trivial developments lacking strategic alignment. Addressed through executive reviews.

SAMPLE CASE 3: INITIAL AI EXPERIMENTATIONS

Industry: Retail

Company Size: Small Business

Business Scenario: A specialty grocery retailer aimed to explore AI powered customer engagement including personalized promotions and conversational chatbots. However, they lacked expertise to evaluate solutions.

AI Solution: They collaborated with a local university AI lab to sponsor an exploratory prototype project. Students gained real-world experience while the retailer accessed cutting-edge perspectives guiding their roadmap.

Outcomes and Impact: The collaboration led to creation of a working conversational bot pilot with strong retention metrics proving viability to scale with vendor solutions.

Implementation Challenges: Ensuring student deliverables met production grade standards. Addressed through enhanced supervisor monitoring.

Research and Benchmarking Tools

To visualize generative AI industry trends effectively in a workshop setting, you can utilize several frameworks and tools. These frameworks help in organizing and presenting information in an engaging and informative manner, making complex data more accessible and understandable. Here are a few frameworks that can be particularly useful:

- SWOT Analysis: This classic framework helps in identifying Strengths, Weaknesses, Opportunities, and Threats related to generative AI in specific industries. It's a straightforward way to break down the current state of generative AI, including its potential for growth and the challenges it faces.

- PESTLE Analysis: This framework examines the Political, Economic, Social, Technological, Legal, and Environmental factors influencing the generative AI landscape. It's particularly useful for understanding external factors that could impact the development and adoption of generative AI.

- Hype Cycle (Gartner-style): The Gartner Hype Cycle is particularly well-suited for technology trends. It provides a graphical representation of the maturity, adoption, and social application of specific technologies. You can create a customized Hype Cycle for generative AI, showing its journey from innovation trigger to the plateau of productivity.

- Technology Adoption Lifecycle: This model, including the diffusion of innovations curve, can help visualize the adoption rate of generative AI. It segments the adoption into categories like Innovators, Early Adopters, Early Majority, Late Majority, and Laggards.

- Value Chain Analysis: This tool can be used to visualize how generative AI adds value at different stages of the industry value chain, from research and development to customer service.

- Mind Maps: Mind mapping is a more free-form but effective way to visualize the various components of the generative AI industry, including key players, technology trends, application areas, and challenges.

- Roadmaps: Technology roadmaps can illustrate the projected development of generative AI over time, showing key milestones, projected advancements, and future goals.

- Data Visualization Tools: Tools like Tableau or Power BI can be used to create dynamic and interactive visualizations of industry data related to generative AI, such as market size, investment trends, patent filings, and more.

- Canvas Models (e.g., Business Model Canvas): These are great for workshops focusing on how businesses can leverage generative AI. They help in visualizing the business aspects of AI implementation, including customer segments, value propositions, channels, and revenue streams.

- Scenario Planning Grids: These are useful for exploring different future scenarios for generative AI, helping participants to think through various possibilities and prepare for a range of potential futures.

In a workshop setting, these frameworks can be used interactively, encouraging participants to add their insights or to use them as a basis for group discussions and brainstorming sessions. Visual aids like posters, digital presentations, or collaborative digital boards (like Miro or Mural) can enhance the engagement and effectiveness of these frameworks.

Generative AI Ideation Canvas

A visualization tool structured as a canvas or table to inspire awareness and kickstart ideation on potential use cases of generative AI. It categorizes examples across:

- Domains: Text generation, Image generation, Video/Animation, Audio generation, Data generation, and Molecule/Material generation, for example.

- Industries/Sectors: Technology, Healthcare & Pharma, Retail & eCommerce, Finance, Manufacturing, Media & Entertainment, Automotive, Energy & Sustainability, Government & Public Sector, Telecom, Logistics & Transportation, Agriculture, and Education.

- Illustrative Use Cases and Examples: Under each domain and industry/sector, provide 3-5 examples of potential generative AI applications.

Here are domain and industry/sector pairs with 3-5 illustrative use case examples of potential generative AI applications for each.

Text Generation

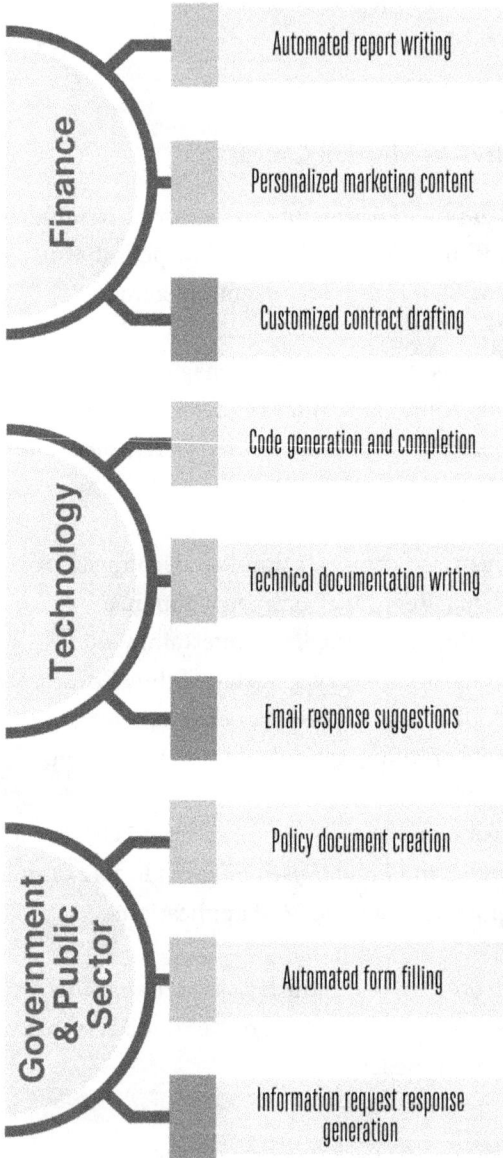

Finance
- Automated report writing
- Personalized marketing content
- Customized contract drafting

Technology
- Code generation and completion
- Technical documentation writing
- Email response suggestions

Government & Public Sector
- Policy document creation
- Automated form filling
- Information request response generation

Image Generation

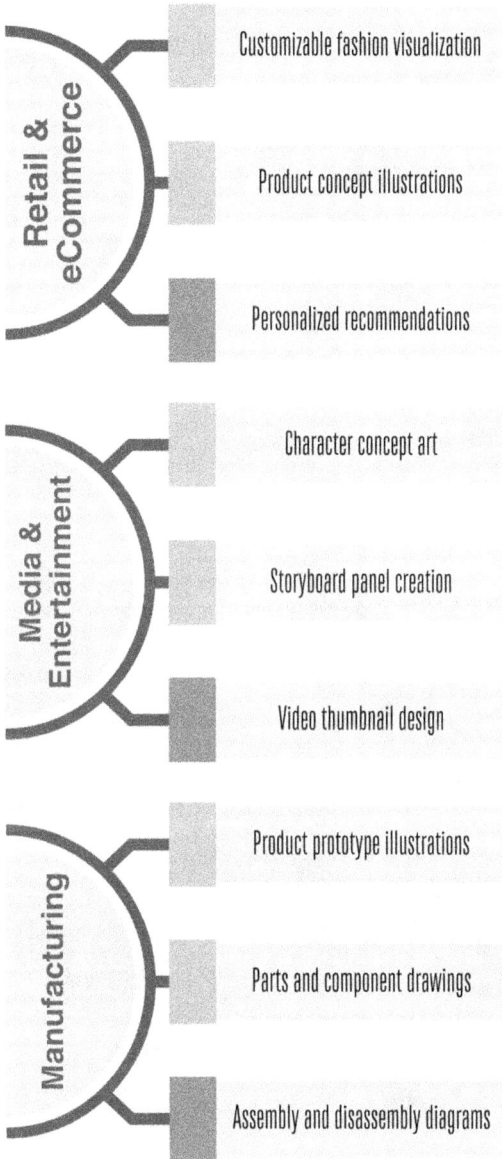

Retail & eCommerce
- Customizable fashion visualization
- Product concept illustrations
- Personalized recommendations

Media & Entertainment
- Character concept art
- Storyboard panel creation
- Video thumbnail design

Manufacturing
- Product prototype illustrations
- Parts and component drawings
- Assembly and disassembly diagrams

Audio Generation

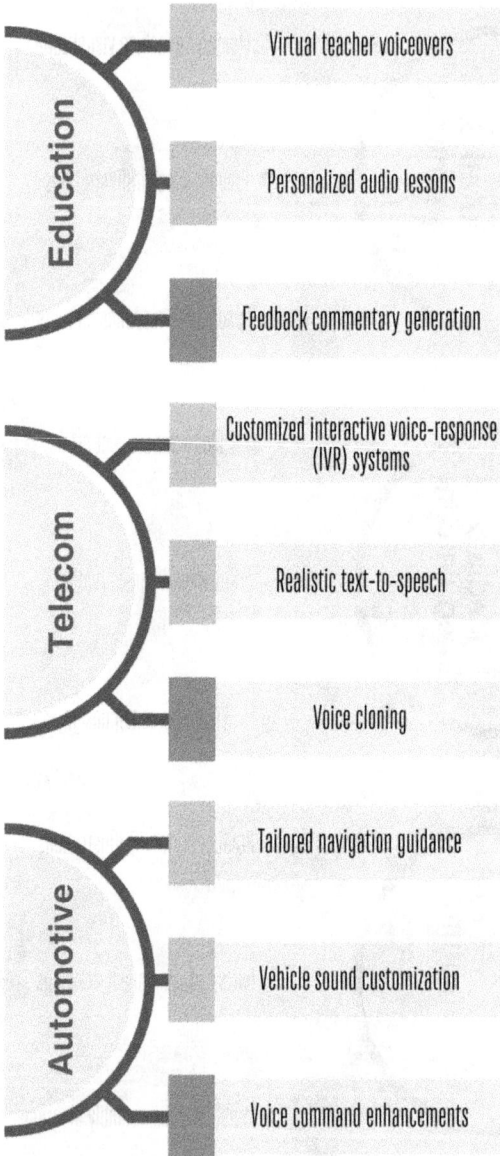

Education
- Virtual teacher voiceovers
- Personalized audio lessons
- Feedback commentary generation

Telecom
- Customized interactive voice-response (IVR) systems
- Realistic text-to-speech
- Voice cloning

Automotive
- Tailored navigation guidance
- Vehicle sound customization
- Voice command enhancements

<u>Data Generation</u>

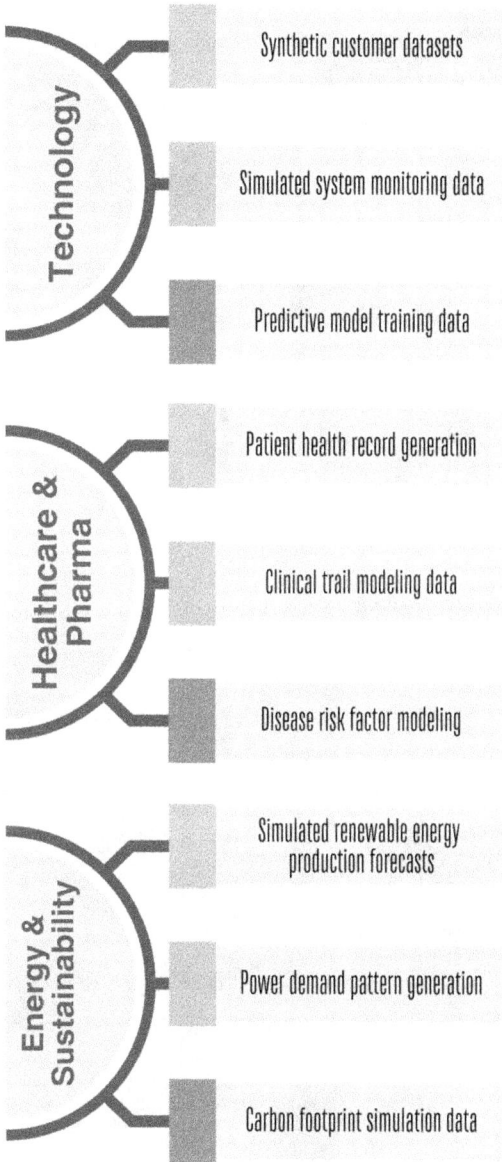

Technology
- Synthetic customer datasets
- Simulated system monitoring data
- Predictive model training data

Healthcare & Pharma
- Patient health record generation
- Clinical trail modeling data
- Disease risk factor modeling

Energy & Sustainability
- Simulated renewable energy production forecasts
- Power demand pattern generation
- Carbon footprint simulation data

<u>Molecular Generation</u>

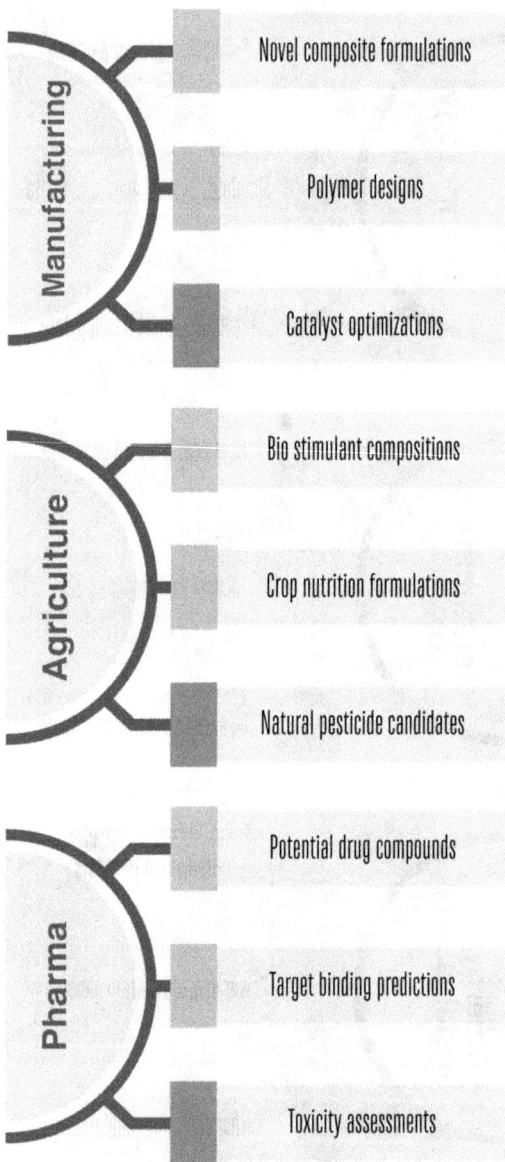

Manufacturing
- Novel composite formulations
- Polymer designs
- Catalyst optimizations

Agriculture
- Bio stimulant compositions
- Crop nutrition formulations
- Natural pesticide candidates

Pharma
- Potential drug compounds
- Target binding predictions
- Toxicity assessments

The examples aim to inspire ideation by demonstrating the wide range of possibilities across domains and sectors. Additional rows can be added for participants to build out further use case ideas specific to their contexts during sessions. References to industry examples or case studies can also be included where relevant.

This canvas structure enables easy scanning while triggering creative thinking on where generative AI can provide value. It can kickstart fruitful ideation workshops or working sessions focused on opportunity exploration.

Relevant Research

Recent research and observations from major consulting firms, technology leaders, and academics highlight the significant impact of generative AI on productivity across various industries:

1. Rapid business adoption: A global survey by McKinsey & Company[2] reveals the rapid adoption of generative AI tools in business functions, with one-third of respondents reporting regular use. This adoption is particularly notable among organizations that already saw significant benefits from traditional AI capabilities. McKinsey's research anticipates significant business disruption, predicting changes to workforce dynamics and substantial industry-specific impacts. They also note a need for policies governing the use of generative AI technologies and mitigating associated risks like inaccuracy.

2. Increase productivity, particularly for routine writing tasks: A MIT study[3] focused on generative AI's impact on worker productivity in tasks like writing cover letters and emails. Findings showed a 40% decrease in task completion time and an 18% rise in output quality when using ChatGPT. The study highlights the potential for generative AI to reduce performance inequality among workers and increase productivity, particularly for routine writing tasks.

3. Productivity focus and readiness gaps: An IBM Study[4] found that nearly half of the CEOs surveyed see productivity as their top priority and are integrating generative AI into their products and services. Despite the enthusiasm, there's concern about data security and bias. The study also uncovers a readiness gap, with fewer executives feeling prepared to adopt generative AI responsibly compared to CEOs. The study emphasizes the belief that organizations with advanced generative AI will gain a competitive edge.

4. Customer support enhancement: A study[5] by Erik Brynjolfsson, Danielle Li, and Lindsey R. Raymond, using data from 5,000 agents at a Fortune 500 software company, demonstrates substantial productivity gains in customer support. The introduction of a generative AI tool led to a 13.8% increase in the number of customer issues resolved per hour. Notably, less experienced and lower-skilled workers saw a productivity boost of 35%. The tool also contributed to reducing agents' communication time per chat by about 9%, handling 14% more chats per hour, and increasing successful resolution rates by 1.3%. Interestingly, customers interacting with AI-assisted agents were less likely to request supervisor help, and agent attrition rates were 8.6% lower.

5. General productivity gains: A compilation of three studies summarized by Nielsen Norman Group[6] indicates an overall increase in business users' productivity by 66% when using generative AI tools like ChatGPT for realistic tasks. The studies involved different user groups—customer service agents, business professionals, and programmers—with significant productivity gains in each group. Customer service agents handled 13.8% more inquiries per hour, business professionals wrote 59% more documents per hour, and programmers coded 126% more projects per week. These results suggest a trend where more cognitively demanding tasks benefit more from AI assistance.

6. Comparison to natural productivity growth: The 66% productivity gains from AI are starkly contrasted with the average labor productivity growth of 1.4% per year in the United States and 0.8% per year in the European Union. These gains from AI equate to several decades of natural productivity growth, highlighting the significant impact of AI on business efficiency.

7. Potential in UX design: Although specific data on UX professionals is limited, it's anticipated that AI could lead to a 100% productivity gain in AI-supported UX tasks. This estimate is based on the complexity of these tasks and the potential for AI to assist in areas like thematic analysis of questionnaire responses.

8. Quality of work and skill gap reduction: The studies also indicate an improvement in the quality of work produced with AI assistance. For instance, in business document writing, the average quality rating improved significantly when composed with AI. Furthermore, generative AI is shown to narrow the skill gap between the best and least talented employees, particularly benefiting those with fewer years of experience or those who spend less time on specific tasks.

9. Accelerated learning: In customer support, agents using AI tools achieved expertise much faster than those without AI support, reducing the time to reach a certain level of productivity from 8 months to just 2 months. This accelerated learning curve underscores the potential of AI in speeding up employee development.

In summary, the latest research and observations from these key players demonstrate the significant and growing impact of generative AI on productivity across various industries. The technology is rapidly being integrated into business functions, offering substantial improvements in efficiency and creativity, while also presenting challenges in implementation, risk management, and ethical considerations.

---

1. Press Release: Gartner Identifies the Top Strategic Technology Trends for 2022.
2. McKinsey survey: The state of AI in 2023: Generative AI's breakout year, August 1, 2023.
3. Study finds ChatGPT boosts worker productivity for some writing tasks by Zach Winn, MIT News Office, July 14, 2023.
4. IBM Study: CEOs Embrace Generative AI as Productivity Jumps to the Top of their Agendas, Jun 27, 2023.
5. Measuring the Productivity Impact of Generative AI, June 2023.
6. AI Improves Employee Productivity by 66%, by Nielsen Norman Group, July 16, 2023.

GENERATIVE AI STRATEGY AND ALIGNMENT

The explosive pace of advancement in generative AI means that organizations risk either missing out on harnessing revolutionary capabilities or investing resources sub-optimally without a coherent strategy guiding implementations.

Effective integration requires alignment to overarching business goals to build stakeholder confidence and focus efforts on high-impact use cases. This strategic grounding also ensures continuity in scaling and evolution amidst a turbulent technological landscape.

This chapter provides guidance on two vital elements for strategic alignment of generative AI initiatives:

1. Mapping generative AI goals to core organizational objectives

2. Prioritizing the most promising applications areas and managing risk

Alignment with Business Goals

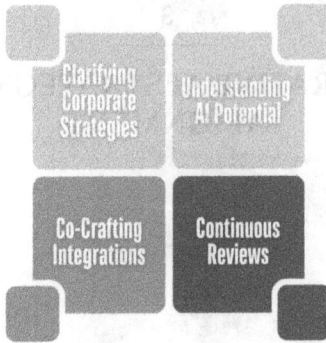

Forging tight linkage between generative AI roadmaps and established organizational strategies including vision, mission and multi-year corporate plans is crucial. Often, companies enthralled by cutting-edge AI hype neglect this basics. The outcomes range from lack of long-term commitment to fragmentation across misaligned sub-projects.

Effective alignment entails:

- Clarifying Corporate Strategies: As strategies evolve, it is vital to have clarity on elements like 2-5 year vision, core mission focus areas, multi-year corporate plan milestones, and short-term business objectives across units.

- Understanding AI Potential: With accelerating R&D across generative AI subdomains ranging from computer vision to natural language, vast new capabilities are emerging continuously. Tracking this landscape spotlights possibilities.

- Co-Crafting Integrations: Joint sessions between business strategy heads and AI leads to systematically map high-potential applications areas against elements of the corporate strategy and establish an integrated implementation sequence synchronizing business and AI milestones over 2-5 year horizons.

- Continuous Reviews: In the rapidly shifting landscape of generative AI, strategies risk becoming stale quickly. Building constant review cycles through AI advisory councils and leadership workshops fosters responsiveness.

Identifying High-Impact Use Cases

With clarity on strategy integration opportunities, a vital next step is to identify and selectively pursue only generative AI application areas offering substantial value.

Promising use cases satisfy criteria like:

- Clear alignment to strategic goals

- Significant magnifying impact on business priorities like revenue boosts, risk reduction or cost optimization

- First-mover or early adopter advantages over competitors

- Relatively more mature underlying AI technology foundations providing confidence in scalability

A combination of discovery approaches helps spot high-potential areas:

- Ideation Workshops: Cross-functional sessions scoped by strategy priorities bringing diverse perspectives across units like R&D, product groups, marketing, risk, and frontline teams.

- Customer Research: Domain experts interviewing B2B clients and end-consumers to identify unmet needs generative AI could address. Allows better demand estimation.

- Market Scanning: Analyzing industry use cases, startup activity, and IP filings to pinpoint areas witnessing generative AI traction globally. Provides signals on viability.

The roadmapping process distills insights from these efforts into a prioritized portfolio of use cases for phased implementation calibrated with business milestones.

Adoption Pathways

Capabilities Analysis

Beyond strategic alignment, a pragmatic adoption roadmap attuned to organizational strengths and constraints is indispensable for generative AI success.

The first step is objective capability analysis assessing areas like data infrastructure maturity, analytics talent availability, access to industry expertise, ethical governance protocols and cultural adaptiveness to change through tools like SWOT, PESTLE etc.

Such audits spotlight preparedness levels shaping adoption pathways.

Stage-Gated Roadmaps

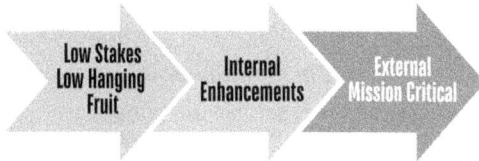

With insights on strengths and gaps, a multi-phase adoption roadmap calibrated to incrementally scale capabilities reduces risks:

- Phase 1—Foundational Use Cases: This entails low stakes/low complexity pilots like generative writing assistants and conversational bots establishing value delivery before investing in complex initiatives.

- Phase 2—Internal Enhancements: Next, improving internal helpdesks, content workflows or forecasting through generative techniques builds credibility within the organization.

- Phase 3—External / Mission Critical AI: Finally, cautiously expanding to high-value customer-impacting solutions like real-time recommendation engines or revenue-generating service enhancements after attaining organizational proficiency.

Conclusion

Strategic alignment and sharp focus safeguard generative AI investments, increasing returns substantially while de-risking initiatives. With pragmatic frameworks to achieve this, companies can transit confidently from experimental phases to integrated enterprise-wide adoption at scale.

Sample Cases in Action

To bring the concepts of this chapter to life and demonstrate their practical application in various industries, I present a series of illustrative use cases. These cases, drawn from a range of sectors and organizational contexts, are designed to help readers envision real-world scenarios. They offer insights into potential opportunities and risks, suggest strategies for mitigation, and guide on evaluating outcomes.

Through these examples, readers can better understand how to apply the principles discussed in this chapter to their own unique situations and challenges.

Sample Case 1: Aligning an AI Center of Excellence

Industry: Automotive

Company Size: Large Enterprise

Business Scenario: A leading automaker sought to launch a new AI Center of Excellence (CoE). Ensuring alignment with overall business objectives was crucial for securing sustained investments.

Solution: The AI CoE leadership conducted joint workshops with division heads identifying use cases to enhance manufacturing optimization, customer personalization and self-driving capabilities—all core strategic priorities.

Outcomes and Impact: The exercises built stakeholder buy-in, culminating in $300 million funding for the CoE aligned to corporate goals. Pilots are underway in the three priority areas.

Implementation Challenges: Securing participation from multiple busy stakeholder groups. Addressed through executive mandate.

Sample Case 2: Generative Design Integration

Industry: Engineering

Company Size: Mid-sized Company

Business Scenario: A manufacturing firm sought to harness generative design AI for parts design optimization. However, ad-hoc experiments lacked focus and measurable value.

Solution: The company instituted rigorous assessment of use cases against criteria covering strategic alignment, cost savings potential and technology maturity. This process prioritized optimizing airflow efficiency for industrial equipment components.

Outcomes and Impact: The optimized generative design solution increased airflow 14% translating to 8% reduced power consumption worth over $2 million in savings annually.

Implementation Challenges: Change management as engineers initially distrusted autonomous tools. Mitigated through extensive upskilling.

Sample Case 3: AI Exploration Task Force

Industry: Financial Services

Company Size: Small Fintech

Business Scenario: A fintech lender recognized opportunities from data-centric AI techniques but lacked structured approaches to identify use cases aligned with their operations optimization goals.

Solution: They instituted a lean cross-functional AI Exploration Task Force including operations leads, technical architects and data scientists chartered with discovering high-impact applications.

Outcomes and Impact: The task force spotted and validated application of natural language generation for automated report writing saving over 200 hours of effort monthly.

Implementation Challenges: Preventing the group from meandering into non-strategic detours. Addressed through executive oversight.

Helpful Tools

Various tools and methodologies can be particularly effective in aiding the strategic alignment process of generative AI within organizations:

- **Balanced Scorecard:** This tool helps in aligning business activities to the vision and strategy of the organization, improving internal and external communications, and monitoring organizational performance against strategic goals.

- **OKRs (Objectives and Key Results):** OKRs are a goal-setting framework used to define and track objectives and their outcomes. They are highly effective in ensuring that the organization's strategy is well-aligned with measurable results.

- **Strategy Maps:** A visual representation tool that articulates the organization's strategic goals and shows the cause-and-effect relationship between them. This can be particularly useful in illustrating how generative AI initiatives align with broader organizational objectives.

- **McKinsey 7S Framework:** This model can be used to analyze organizational effectiveness and align internal aspects with external environmental changes. It considers seven internal aspects of an organization: strategy, structure, systems, shared values, skills, style, and staff.

- **Porter's Five Forces:** This framework is useful for analyzing the competitive environment and understanding where generative AI can be leveraged to gain a competitive advantage or mitigate threats.

- **Ansoff Matrix:** This strategic planning tool provides a framework to help executives, senior managers, and marketers devise strategies for future growth, including the integration of new technologies like generative AI.

- **Change Impact Analysis Tools:** These tools help in assessing the potential impact of changes brought by generative AI. They can identify areas of the business that will be most affected and help in planning accordingly.

- **Gap Analysis Tools:** These tools are used to compare actual performance with potential or desired performance. This is particularly useful in strategic planning for generative AI to identify where the organization currently stands and where it aims to be.

- **Stakeholder Analysis and Management Tools:** Tools like stakeholder maps or matrices help in identifying and managing the expectations and influence of various stakeholders in the process of aligning AI strategy.

- **Risk Management Tools:** Tools like risk matrices or risk registers are essential in strategic planning, especially to identify, assess, and mitigate risks associated with implementing generative AI.

Each of these tools can be used to support different aspects of strategic alignment, from setting and tracking goals to analyzing internal and external environments, managing change, and assessing risks and impacts. They can be particularly helpful in workshops or strategic planning sessions, providing structured approaches to aligning generative AI initiatives with broader organizational objectives.

4

TALENT AND SKILLS DEVELOPMENT

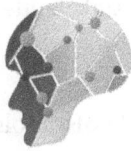

As generative AI capabilities rapidly mature, one of the biggest constraints organizations face is talent. Demand for specialized AI skills far outstrips supply globally. Building a robust generative AI workforce calls for strategic interventions spanning hiring, training and re-skilling.

This chapter provides guidance across two key areas:

1. Building in-house capabilities through training programs to equip existing employees to effectively adopt generative AI.

2. Attracting and retaining external generative AI talent through targeted employer branding.

Upskilling Programs

While hiring laterally has time and cost overheads, uplifting capabilities internally allows leveraging contextual and domain expertise within the organization.

Tailored training programs built on rigorous skills assessment address gaps effectively. Blended formats like live workshops, online modules, coached projects and self-learning leverage technology and scale. Google's machine learning crash courses and IBM's Digital Badge programs exemplify such models.

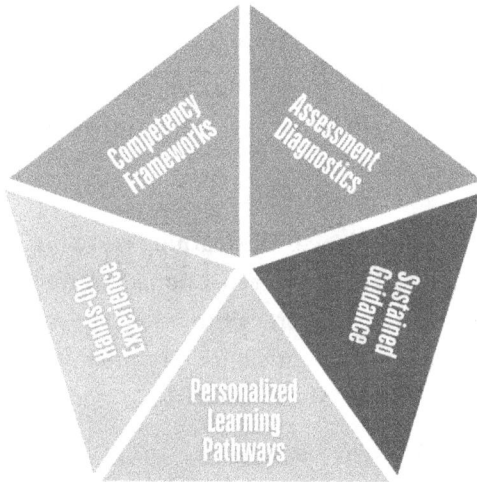

Key elements for designing impactful upskilling include:

- Competency Frameworks: Distinct from generic AI awareness, these outline technical, business and soft skills required for roles working directly with generative AI like data analysts, application engineers and program managers.

- Assessment Diagnostics: AI skills being nascent and fast-evolving, diagnostics allow objective measurement of proficiency levels and readiness for new responsibilities across competencies.

- Personalized Learning Pathways: Based on diagnostics and role contexts, these enable employees to systematically build capabilities through prescribed learning flows comprising different modes like online content, mentor guidance and purpose-built challenges.

- Hands-On Experience: No training hits home as much as hands-on experimentation. From sandbox environments to hackathons and guided commercial pilots, concrete application cement concepts effectively.

- Sustained Guidance: The learning curve for generative AI is often steep. Connecting trainees to mentors, online expert communities and coaching circles for ongoing assistance is invaluable.

Recruiting AI Talent

While robust internal capacity is crucial, the fact remains that generative AI as an emerging field requires lateral hires with specialized expertise. Attracting such talent poses stiff competition.

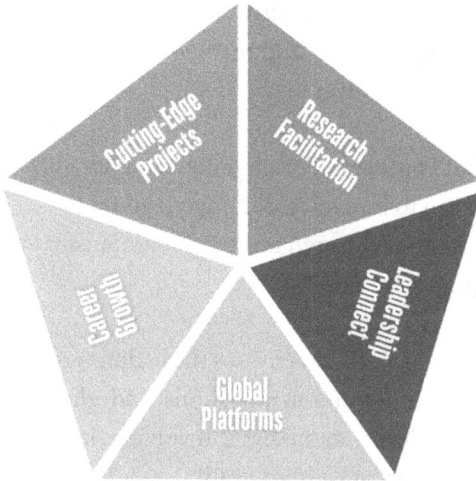

Companies must strengthen their talent brand specifically among generative AI communities by highlighting compelling reasons for experts to join them over peers or remain in academia including:

- Cutting-Edge Projects: Showcasing bleeding-edge application challengesBeing undertaken for clients or internal needs provides technologists exposure unattainable elsewhere.

- Research Facilitation: Guaranteed time, budgets and data access for pursuing research interests piques scientific talent. Co-branding papers helps spread brand affiliation.

- Leadership Connect: Direct access to executive leaders responsible for generative AI direction establishes pivotal influence potential.

- Career Growth: Lucrative career evolution paths spanning technical and managerial ladder steps reassure high-flyers.

- Global Platforms: For larger enterprises, the platform to impact wider organizational applications through international transfers magnifies attraction.

Targeted outreach then activates these differentiators like AI expert talks, university symposiums and global generative AI events. Referral campaigns involving leading affiliated scientists further boost brand pull.

Conclusion

The combination of sustained upskilling drive and magnetic employer brand provides a potent formula for meeting generative AI talent needs as organizations gear up for increasing adoption.

Sample Cases in Action

To bring the concepts of this chapter to life and demonstrate their practical application in various industries, I present a series of illustrative use cases. These cases, drawn from a range of sectors and organizational contexts, are designed to help readers envision real-world scenarios. They offer insights into potential opportunities and risks, suggest strategies for mitigation, and guide on evaluating outcomes. Through these examples, readers can better understand how to apply the principles discussed in this chapter to their own unique situations and challenges.

Sample Case 1: Designing an AI Reskilling Curriculum

Industry: Technology Services

Company Size: Large Enterprise

Business Scenario: A major IT services company sought to reskill its 100,000 strong workforce to meet surging client demand for AI solutions expertise.

Solution: They developed an AI competency framework, assessed talent strengths/gaps via analytics and created personalized learning pathways comprising online courses, mentor projects and certification incentives guiding reskilling.

Outcomes and Impact: Over 20,000 employees completed AI reskilling in 18 months. This expanded their service delivery capacity sharply, unlocking $15 million in incremental AI services revenue.

Implementation Challenges: Motivating staff by clarifying career evolution pathways post-reskilling.

Sample Case 2: University AI Talent Pipeline

Industry: Consumer Goods

Company Size: Large Conglomerate

Business Scenario: A leading FMCG brand sought to tap specialized AI/ML talent for its analytics and supply chain automation initiatives. However, they found it challenging to attract experts away from tech firms.

Solution: They instituted university partnerships facilitating graduate internships and research collaborations. This cultivated an early talent pipeline with graduates coveting the company's cutting-edge initiatives.

Outcomes and Impact: Pipeline strengthened attracting top-tier data science graduates for full-time positions. Interns also catalyzed innovations like optimizing promotional budget allocation.

Implementation Challenges: Offers getting outbid by tech giants. Mitigating through exclusive project participation incentives.

SAMPLE CASE 3: FOCUSED AI TEAM SCALING

Industry: Banking Services

Company: Mid-sized Firm

Business Scenario: A banking process outsourcing firm identified document data extraction as a use case to minimize manual effort. However, they lacked specialized AI talent to implement solutions.

Solution: They competitively hired scarce machine learning expert talent through focused brand-building among university communities demonstrating fascinating text analytics challenges surpassing typical fintech firms.

Outcomes and Impact: The initiative expanded their document digitization capacity 3X, bolstering new contract wins worth over $5 million.

Implementation Challenges: Retaining new AI experts long-term. Addressing via engaging research support and skills retraining scope.

―――――

Helpful Tools

A Generative AI Competency Framework can be laid out in a table format, categorizing key competencies into different areas, helping to identify and develop the skills necessary for working effectively with generative AI technologies. Here's a simplified version of what such a framework might look like:

Key Competencies	Description/Examples
Technical Skills	
Machine Learning and AI Algorithms	Understanding of AI and ML algorithms, including neural networks, GANs, reinforcement learning, etc.
Data Engineering	Skills in data preprocessing, cleansing, and engineering for AI model training and deployment.
Software Development	Proficiency in programming languages like Python, R, and frameworks like TensorFlow, PyTorch.
Applied Skills	
Problem-Solving with AI	Ability to apply AI solutions to real-world problems, understanding AI's capabilities and limitations.
AI Model Development and Management	Skills in developing, training, and managing AI models, including understanding of model lifecycle management.
Business Acumen	
Strategic Thinking	Understanding how generative AI can align with and support business goals and strategies.
Industry Knowledge	Awareness of industry-specific applications and implications of AI.
Interpersonal Skills	
Communication and Collaboration	Ability to effectively communicate AI concepts and collaborate with non-technical teams.
Ethical and Social Awareness	Understanding of ethical considerations and societal impacts of AI.
Adaptive Skills	
Continuous Learning	Willingness and ability to continually learn and adapt to the rapidly evolving field of AI.
Creativity and Innovation	Ability to think creatively and propose innovative solutions using AI.
Leadership and Governance	
AI Strategy and Policy Development	Skills in developing AI strategies and policies that align with organizational goals.
Risk Management and Compliance	Understanding of the risks associated with AI applications, including data privacy and security, and compliance with regulations.
Stakeholder Management	Ability to manage expectations and engage stakeholders effectively in AI projects.

5

RISK MANAGEMENT AND ETHICS

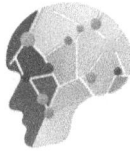

As organizations harness the power of generative AI, they also inherit complex risks ranging from system biases to data privacy threats. Without earnest efforts to address these proactively, initiatives risk reputational backlash or even legal non-compliance.

This chapter highlights two major risk areas needing governance—biases leading to unreliable or harmful AI behaviors and data privacy concerns—alongside strategies to manage them responsibly.

Bias and Reliability

Generative AI systems intrinsically reflect biases and inaccuracies from their training data. Real-world data often incorporates societal prejudices around race, gender and more leading models to perpetuate harm through exclusions or problematic content.

Issues also stem from the inherent randomness within generative algorithms producing erratic outputs occasion-

ally or training set distortions yielding plausibly fluent but
logically incorrect text or imagery.

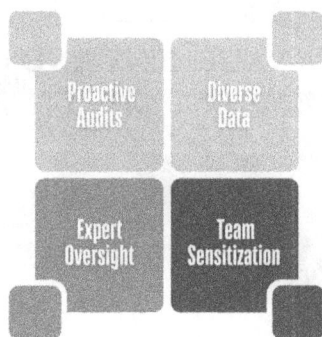

Robust governance is indispensable to address biases and
reliability gaps. Leading practices involve:

- Proactive Audits: Instead of waiting for troublesome
  incidents, continuously monitor systems and
  outputs using bias scanning tools. Human-in-the-
  loop analysis also helps spot anomalies early
  despite fluent facades.

- Diverse Data: Actively source training data to
  minimize representation skew across demographic
  groups. Frequent retraining with updated datasets
  curtails data drift issues.

- Expert Oversight: Establish review processes
  involving bias specialists and domain experts for
  generative model approvals before deployment.
  Continued guidance bridges gaps pending
  technological maturity.

- Team Sensitization: Beyond tools, much hinges on human judgment. Training data team members to detect biases and providing simple mitigation protocols like risk checklists prevents negligence.

Data Privacy and Security

Central to generative AI is data—the lifeblood fueling capabilities. Vast datasets integrate insights from people ranging from clinical trial volunteers to social media users or retail browser behaviors to train commercial systems.

Stringent data governance combining legal compliance and ethical practices provides the license to operate such AI responsibly.

Key focus areas are:

- Anonymization Standards: As models ingest datasets with personal information, robust protocols must irreversibly de-identify records securing identities.

- Access Controls: Multi-layered authentication, monitoring and privileged access measures combat external and insider data theft threats. Aligns to regulations like HIPAA.

- Use-case Scrutiny: Ensure applications amount to consented use-cases with clear service provider contracts. Helps catch regulatory gaps or infringements requiring course correction.

- Transparency: Apart from mandates in regulations like EU's GDPR and AI Act, being transparent on data usage issues like aggregation, retention policies or model designs even without legal needs builds trust.

Conclusion

Earning stakeholder confidence in AI through earnest and transparent efforts to tackle some of its most problematic pitfalls culminates ultimately in its safe adoption and sustained impact.

Sample Cases in Action

To bring the concepts of this chapter to life and demonstrate their practical application in various industries, I present a series of illustrative use cases. These cases, drawn from a range of sectors and organizational contexts, are designed to help readers envision real-world scenarios. They offer insights into potential opportunities and risks, suggest strategies for mitigation, and guide on evaluating outcomes. Through these examples, readers can better understand how to apply the principles discussed in this chapter to their own unique situations and challenges.

Sample Case 1: Proactive AI Ethics Reviews

Industry: Public Sector

Organization Size: Government Agency

Business Scenario: A healthcare agency sought to pilot AI for policy design. However, biased data could skew social support programs, necessitating reviews.

Solution: Cross-functional teams scrutinized data composition, algorithms and use cases through equality impact assessments. Corrective tweaks were mandated addressing gender, age and accessibility biases identified proactively.

Outcomes and Impact: The exercises averted risks of exclusion or discrimination while boosting public trust and participation essential for large-scale rollout.

Implementation Challenges: Building understanding within teams on biases going beyond technical accuracy.

Sample Case 2: Customer Data Privacy Optimization

Industry: Retail

Company Size: Large Organization

Business Scenario: An ecommerce major capturing vast shopper data for recommendations had fragmented privacy practices across markets. This inhibited harnessing data for global AI needs.

Solution: They instituted centralized anonymization protocols, access controls and retention policies exceeding legal minimums to catalyze customer trust and enable accountable data sharing across borders.

Outcomes and Impact: Published voluntary transparency reports built brand reputation while accelerating launch of personalized product recommendation algorithms worth $100 million in incremental revenue.

Implementation Challenges: Coordinating streamlined policies across regional teams. Resolved via executive directive.

SAMPLE CASE 3: AI ALGORITHM BIAS MITIGATION

Industry: Banking

Company Size: Mid-sized Firm

Business Scenario: A digital bank building automated credit approval models using alternative data ran into bias issues skewing application decisions unfairly.

Solution: They implemented layers of oversight including bias detection tools, expert auditors and external reviews to catch and correct distortions through iterative data and model updates.

Outcomes and Impact: The governance structures increased model fairness and accuracy, catalyzing approvals worth $2 million in previously missed income.

Implementation Challenges: Quantifying complex bias mitigation metrics. Improved via research collaboration.

Responsible AI Framework

A Responsible AI Framework serves as a navigational tool, ensuring that while we harness the power of AI to innovate and solve complex problems, we remain steadfast in our commitment to ethical principles and societal well-being.

This section of the book delves into the various components of a Responsible AI Framework. Each component is designed to address specific aspects of AI ethics and governance—from ensuring fairness and transparency to managing data privacy and security risks. The framework is not just a theoretical construct; it is a practical guide, complete with actionable strategies and real-world applications.

As we explore this framework together, I invite readers to consider not just the "how" of AI implementation, but the "why" and "what for". It is an invitation to engage in a deeper conversation about the role of AI in our society and the legacy we wish to create through its use. Our journey through the Responsible AI Framework is more than a pathway to compliance or risk management; it is a commitment to a future where technology amplifies human potential and operates within the bounds of ethical responsibility.

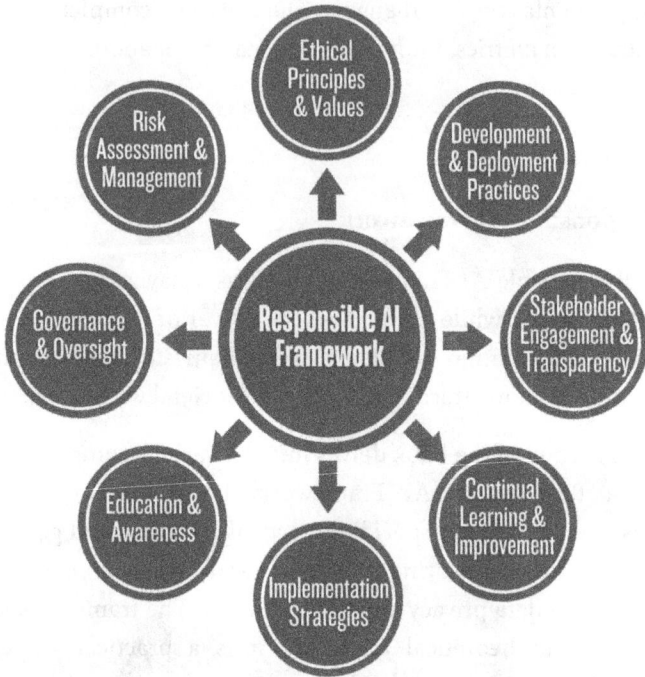

1. Ethical Principles and Values

- **Transparency:** Ensure that AI systems are transparent in their operations and decision-making processes.

- **Fairness and Non-Discrimination:** Address biases in data and algorithms to prevent discrimination based on race, gender, or other societal biases.

- **Accountability:** Assign clear responsibility for AI system behavior and outcomes.

- **Privacy and Data Governance:** Protect user data privacy, ensure data security, and manage data ethically. Emphasize consent-based data usage and respect for intellectual property rights.

2. Risk Assessment and Management

- **Bias and Reliability Analysis:** Regularly conduct assessments to identify and mitigate biases in AI systems.

- **Impact Assessment:** Evaluate the potential social, ethical, and environmental impact of AI systems before and after deployment.

- **Security and Privacy Risks:** Assess and implement measures to protect AI systems from data breaches and misuse.

3. Development and Deployment Practices

- **Inclusive Design and Development:** Involve diverse groups in AI design and development to ensure wide-ranging perspectives and needs are considered.

- **Robust and Reliable AI Systems:** Develop AI systems that are robust, reliable, and capable of handling errors or unexpected situations.

- **Continuous Monitoring:** Implement ongoing monitoring of AI systems to ensure performance and ethics are maintained post-deployment.

4. Governance and Oversight

- **Regulatory Compliance:** Adhere to all relevant local and international AI regulations and standards.

- **Internal Ethics Boards:** Establish an internal ethics board or committee to oversee AI initiatives and ensure adherence to ethical standards.

- **External Audits and Certifications:** Regularly conduct external audits and pursue certifications for AI systems to maintain transparency and trust.

5. Stakeholder Engagement and Transparency

- **Stakeholder Communication:** Maintain open lines of communication with stakeholders regarding AI development and its impacts.

- **Public Reporting and Disclosure:** Publicly disclose information about AI systems' capabilities, limitations, and performance metrics.

6. Education and Awareness

- **Training Programs:** Provide training for employees and stakeholders on ethical AI use and awareness.

- **Public Education Initiatives:** Engage in or support public education initiatives to raise awareness about AI ethics and responsible use.

7. Continual Learning and Improvement

- **Feedback Loops:** Establish feedback mechanisms to continuously learn from AI operations and stakeholder input.

- **Adaptation to Emerging Technologies and Norms:** Regularly update the framework to reflect new technological advancements and evolving societal norms.

8. Implementation Strategies

- **Toolkits and Resources:** Leverage existing toolkits and resources from credible sources for implementing various components of the framework.

- **Partnerships:** Collaborate with academic, industry, and regulatory bodies to stay informed and align with best practices.

By developing a Responsible AI Framework, organizations can ensure that their use of generative AI aligns with ethical standards and best practices, while effectively managing associated risks.

Best Practices for Establishing AI Ethics Boards

Establishing AI Ethics Boards is a vital step in ensuring responsible AI use, but it should not be the sole line of defense. Instead, these boards should function as part of a multi-tiered approach to AI ethics within an organization. Here, I outline best practices for establishing AI Ethics Boards.

1. Layered Approach to AI Ethics:

- Multiple Lines of Defense: AI Ethics Boards should be the last line of defense in a multi-layered ethical oversight structure. This structure should start with ethical design principles embedded in the AI development process, followed by internal review teams and compliance checks before escalating to the Ethics Board.

- Responsibility at All Levels: Every team member, from developers to executives, should be trained and held accountable for ethical considerations in their work. Ethical AI is a collective responsibility, not just that of the Ethics Board.

2. Diverse and Multidisciplinary Membership:

- Inclusive Composition: Include members from diverse backgrounds and expertise on the board for a holistic view of ethical issues.

- Broad Representation: Ensure representation from various impacted groups to address diverse ethical concerns effectively.

3. Clear Mandate and Objectives:

- Defined Role: Establish a clear, detailed mandate outlining the board's role, responsibilities, and decision-making limits.

- Focused Areas: Clearly define the board's areas of focus, including critical ethical issues like data privacy and algorithmic bias.

4. Autonomy and Authority:

- Operational Independence: Ensure the board operates independently from the company's commercial interests.

- Empowered Decision-making: Grant the board authority to make impactful recommendations or decisions on AI ethics.

5. Access to Information and Expertise:

- Unrestricted Access: Guarantee access to necessary information about AI projects for informed decision-making.

- External Consultations: Facilitate engagement with external experts for broader perspectives.

6. Transparent and Accountable Operations:

- Open Communication: Implement transparent reporting of the board's activities and decisions.

- Organizational Accountability: Develop processes for the entire organization to be accountable for upholding AI ethics, not just the board.

7. Regular Review and Adaptation:

- Evolving Practices: Regularly update the board's practices to keep pace with changing AI technologies and societal standards.

- Feedback Mechanisms: Establish feedback loops for continuous learning from the board's recommendations.

8. Education and Organizational Culture:

- Continuous Learning: Provide ongoing education for board members and all organizational members on AI developments and ethical considerations.

- Ethical Culture: Foster an organizational culture where ethical AI is a shared value and responsibility.

9. Collaboration and Stakeholder Engagement:

- Collaborative Networks: Work with ethics boards from various sectors to exchange insights.

- Inclusive Dialogue: Actively engage with diverse stakeholders, including those potentially impacted by AI.

10. Policy Development and Implementation Support:

- Guideline Formulation: Assist in creating ethical AI policies and guidelines.

- Practical Implementation: Advise on the effective implementation of these policies across the organization.

11. Evaluation and Organizational Impact Assessment:

- Effectiveness Assessment: Continuously evaluate the impact of the board's work and the organization's adherence to AI ethics.

- Impact Reporting: Document how AI ethics practices influence AI development and use within the entire organization.

In conclusion, while AI Ethics Boards are crucial in guiding ethical AI practices, they should be part of a broader, organization-wide commitment to AI ethics. This approach ensures that ethical considerations are integrated at every stage of AI development and use, creating a culture of responsibility and accountability across the organization.

Counterpoint: Responsible AI Perspectives

While generative AI promises immense benefits, concerns on detrimental impacts ranging from encoded biases to environmental consequences continue arising, necessitating earnest redressal.

Providing vital direction are trailblazing researchers[1] cautioning organizations against technological optimism sans accountability.

I summarize crucial perspectives:

- Key risks from uncontrolled AI progress include marginalization, inequality, authoritarian oppression of minorities facilitated by generative systems. Solutions involve building institutional consciousness on socio-technical issues, cooperation not control of communities affected, and embracing flexibility to meet varied locales' needs.

- Language AI risks losing sight of people's multifaceted identities amid chasing technical metrics like accuracy. Real people and how they are affected should anchor system development. This demands interdisciplinary teams spanning policy, social science and humanities to complement technical specializations coupled with proactive impact assessment.

- The AI field lacks urgency in identifying and measuring types of model harm rigorously before deployment. Safety cannot be an afterthought. Methodical issue mapping, building mathematical understandings of risks like negative side effects and safety-aware design practices provide pathways.

- Like medicine's credo of "first, do no harm", AI demands precautionary principle-based approaches as risks span erosion of empathy, mental health issues and environmental damage. Collective deliberation on beneficial visions for society and human flourishing should guide technological choices judicious of risks and tradeoffs.

These perspectives call for profound reflection on AI's unintended damages and to situate technical build decisions within deeper dialogue on ethical responsibilities steering innovation trajectories wisely.

AI Governance and Policy Landscape

As generative AI capabilities advance, policymakers globally grapple with complex challenges around managing risks as varied as biased impacts, misinformation threats, and even environmental consequences.

Understanding the emergent regulatory landscape is vital for organizations to track compliance needs and contribute constructively to governance evolution centered on the public good. Examples from two major regions:

European Union

The EU unveiled pioneering AI regulations in 2021 spanning mandatory transparency rules, risk-based classifications and fines up to €30 million for non-compliance. Specific focus areas include:

- Banning certain AI uses deemed socially dangerous
- Restricting high-risk applications like medical diagnosis algorithms
- Mandating human oversight for key decisions
- Requiring technical documentation and algorithm auditing

The comprehensive laws govern AI use in the EU marketplace while also influencing global norms.

UNITED STATES

US policy action on AI has gained momentum with several legislative proposals from banning questionable uses to establishing ethical norms. Two key recent milestones include:

- National AI Research Resource Task Force report emphasizing responsible development and use.
- Executive order directing federal agencies to partake in international initiatives promoting trustworthy AI adoption.

But partisan disagreements have stalled enacting overarching federal laws so far.

THE ROAD AHEAD

As generative AI capacities grow more disruptive, managing risks through governance is imperative. Organizations must track evolving regulations proactively while investing earnestly in ethical internal practices exceeding mere compliance.

Responsible generative AI aligning innovations to societal needs relies on collective deliberation. We must participate actively today to shape policies securing equitable, safe and beneficial technological progress for all global citizens tomorrow.

---

1. Prof. Timnit Gebru, Dr. Emily Bender, Dr. Margaret Mitchell, Dr. Alex Hannah, the Dair Institute for AI Responsibility Research, and the AI Now Institute.

6

TECHNOLOGY INTEGRATION AND DEVELOPMENT

Harnessing generative AI's potential needs setting up robust technology foundations within the organization. Unlike other enterprise software, AI systems make exceptionally heavy infrastructural demands. Choosing between custom-built or readymade solutions also poses complex trade-offs.

This chapter offers guidance across two key technology integration challenges:

1. Assessing and upgrading IT infrastructure readiness

2. Optimizing make vs buy decisions for AI solutions

IT and Infrastructure Readiness

Being data-driven, generative AI reliance on high-performance computing capabilities is multiples beyond conventional workloads. Without provisioning adequate data storage, computing power and software environments, productivity sharply suffers.

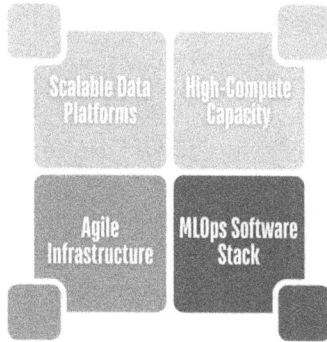

Key focus areas for boosting readiness span:

- Scalable Data Platforms: Assembling vast, multi-modal datasets for model development requires specialized data lakes supporting immense volumes with accelerated access and diverse integration capabilities.

- High-Compute Capacity: Training and running deep learning models demands extensive graphical and tensor processing hardware alongside facile orchestration for cost efficiency through cloud or on-premise high-performance computing infrastructure.

- Agile Infrastructure: Containerization, microservices and devops culture foster rapid iteration crucial for customization agility in contrast to inflexible on-premise systems.

- MLOps Software Stack: To transition experiments into production-grade flows, MLOps layers automate repetitive tasks, standardize pipelines, facilitate cross-team collaboration and enable model monitoring.

Customization vs Off-the-Shelf Solutions

While terminals like OpenAI's DALL-E visually demonstrate generative AI's powers, commercializing applications requires training models on proprietary datasets reflecting unique needs.

Organizations face trade-offs choosing between custom development or licensing readymade solutions as AI-as-a-service then fine-tuning further.

Criteria for optimization span:

- Strategic Differentiation: For some use cases like demand forecasting, off-the-shelf solutions suffice. In differentiated areas like synthetic patient data, custom models are indispensable.

- Implementation Timelines: Licensed models speed up deployment but need rigors of data integration, tuning cycles and infrastructure upgrades just like custom builds. The delta may be marginal.

- Talent Availability: Developing in-house is contingent on securing specialized skill sets. Faster launches may need procurement till organic talent ramps up.

- Total Cost of Ownership: Subscription costs accumulate long-term while amortizing custom development expenses lowers total cost over 5+ year horizons.

Based on use case assessments, blended approaches using custom modules for core IP-sensitive tasks while leveraging pre-trained or cloud-hosted baseline models aid optimization.

Conclusion

With mounting generative AI adoption, the strategies above provide a blueprint for erecting robust technological foundations while optimizing make-vs-buy decisions allowing organizations to extract full value.

Sample Cases in Action

To bring the concepts of this chapter to life and demonstrate their practical application in various industries, I present a series of illustrative use cases. These cases, drawn from a range of sectors and organizational contexts, are designed to help readers envision real-world scenarios. They offer insights into potential opportunities and risks, suggest strategies for mitigation, and guide on evaluating outcomes. Through these examples, readers can better understand how to apply the principles discussed in this chapter to their own unique situations and challenges.

SAMPLE CASE 1: SCALABLE AI INFRASTRUCTURE MODERNIZATION

Industry: Media & Entertainment

Company Size: Large Enterprise

Business Scenario: An animation studio pioneering AI-based content creation faced severe infrastructure limitations hindering experimentation and scale.

Solution: They invested in scalable cloud data lakes, high-performance GPU clusters, containerized model development workflows and MLOps—boosting flexibility for accelerated generative content innovation.

Outcomes and Impact: The revamped environment fast-tracked experiments and complex model iterations catalyzing breakthroughs like real-time 3D animation rendering. Efficiency gains delivered $50+ million savings over 5 years.

Implementation Challenges: Change management as legacy on-premise infrastructure teams resisted cloud shift. Mitigated via reskilling programs.

SAMPLE CASE 2: HYBRID AI SOLUTION OPTIMIZATION

Industry: Computer Software

Company Size: Mid-sized Firm

Business Scenario: An enterprise software provider sought speech and language functionality for contextual search capabilities. Fully custom development was cost-prohibitive despite strategic value.

Solution: They integrated accessible vendor-based speech recognition while customizing downstream natural language understanding for nuanced queries across complex product documentation.

Outcomes and Impact: The hybrid AI solution enhanced search relevance by over 35%, contributing an estimated $3 million revenue bump from improved customer experience.

Implementation Challenges: Seamless integration challenges across disparate components. Resolved via extensive API expansion.

Sample Case 3: Exploring Pre-Built Generative AI

Industry: Management Consulting

Company Size: Small Firm

Business Scenario: A niche advisory firm wanted to harness generative writing tools to boost content productivity but lacked AI expertise. Developing custom solutions however was cost-prohibitive.

Solution: They piloted readily available industrial-grade generative writing software using unique client data and scenarios to create tailored thought leadership efficiently after light tuning.

Outcomes and Impact: Leveraging accessible readymade AI unlocked a 5X increase in high-quality original perspectives shared externally, elevating brand authority.

Implementation Challenges: Perceived quality gap of software-generated content versus fully manual authoring. Mitigated through rigorous multi-stage review workflows.

Generative AI Buy vs. Build Questionnaire

Here's a set of questions designed to guide organizations in deciding whether to purchase off-the-shelf generative AI solutions or develop them in-house. This questionnaire is structured to cover various critical aspects, from strategic alignment to technical feasibility and cost considerations.

Strategic Alignment:

- How does the generative AI solution align with your organization's long-term strategic goals?

- Will the solution provide a competitive advantage that aligns with your business objectives?

Core Competencies and Capabilities:

- Does your organization have the in-house expertise required to build a generative AI solution?

- Are there existing technologies within your organization that can be leveraged for building a generative AI solution?

Cost Considerations:

- What is your budget for the generative AI initiative, and how does it compare with the estimated costs of building vs. buying?

- Have you considered the long-term costs associated with maintenance, updates, and scalability for both options?

Time to Market:

- What is your timeframe for deploying the generative AI solution?

- How does the time to develop a custom solution compare with the implementation time for an off-the-shelf solution?

Customization Needs:

- Does your use case require a highly customized solution that off-the-shelf products cannot adequately provide?

- How critical is it for the solution to be tailored to specific workflows or data types unique to your organization?

Data Privacy and Security:

- What are your data privacy and security requirements, and can an off-the-shelf solution meet these requirements?

- Does building in-house provide more control over data privacy and security concerns?

Vendor Ecosystem:

- Are there reliable vendors in the market offering generative AI solutions that meet your needs?

- How do the support structures, reliability, and track records of these vendors compare?

Future Scalability and Flexibility:

- Will the chosen approach (buy or build) be scalable and flexible enough to accommodate future growth and technological changes?

- How easy is it to update or modify the solution as your needs evolve?

Integration with Existing Systems:

- How will the chosen solution integrate with your existing systems and infrastructure?

- Are there potential technical challenges in integration that might favor one approach over the other?

These questions are designed to provoke thoughtful consideration of the various factors involved in the buy vs. build decision, specifically tailored to the unique challenges and opportunities of generative AI technology. The responses can help organizations form a comprehensive view of their needs, capabilities, and constraints, guiding them toward the most appropriate decision for their specific context.

7

IMPLEMENTATION AND SCALING

The most crucial leap for generative AI success is crossing the chasm from exciting proofs-of-concept or prototypes into full-fledged real-world implementations delivering concrete business gains. This necessitates structured scaling pathways centered on validated pilot projects.

This chapter discusses recommended approaches for:

1. Launching pilots to test generative AI viability before commitment to scale

2. Expanding validated applications through gradual, metrics-driven scaling

Pilot Testing

Pilots are invaluable for generative AI solutions as they verify value delivery addressing genuine business needs on restricted data samples before large-scale engineering investments.

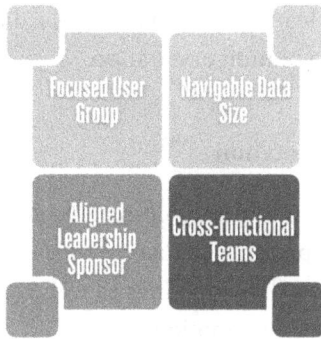

Winning qualities for strong pilot candidates are:

- Focused User Group: Solutions with clearly delineated beneficiary profiles allow accurate evaluation of adoption readiness from user feedback.

- Navigable Data Size: Pilots necessitate slicing production data volume across parameters like time range, geography or customer cohort for agile iteration. AI teams may otherwise drown in complexity.

- Aligned Leadership Sponsor: Having an executive champion invest budget and attention to pilot launches and metrics tracking ensures serious commitment levels for scale-up conditional to positive results.

- Cross-functional Teams: Unlike generic software pilots relying on technical specialists, AI testing gains tremendously from diverse viewpoints. User-experience leads, program managers, data analysts all improve solution orientation.

Scaling Implementation

Transitioning from pilots to full-scale generative AI production involves expanding across technical dimensions like infrastructure and data pipelines as well as cultural change management for user adoption.

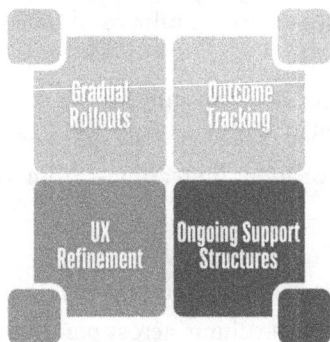

Prudent pathways entail:

- Gradual Rollouts: Stage release across incremental user segments for manageable iterations allowing technology and change absorption. Periodic milestone reviews help course correct.

- Outcome Tracking: Keep quantifiable metrics aligned to original AI motivation like risk reduction targets, conversion lifts or churn containment front and center when calibrating rollout pace.

- UX Refinement: Generative AI brings new paradigms. Continued user research spotlighting evolving interaction pain points necessary for design tweaks, even with interfaces validated during pilots.

- Ongoing Support Structures: Users need assistance adapting workflows or upskilling abilities to fully utilize new AI capabilities implemented in their contexts. Technical helpdesks, chatbots and micro-learning aid adoption.

Conclusion

With rigorous piloting and structured scaling policies tailored to address generative AI's multifaceted user impact and technology challenges, organizations can smoothly transition promising solutions to deliver their full, brink scale potential.

Sample Cases in Action

To bring the concepts of this chapter to life and demonstrate their practical application in various industries, I present a series of illustrative use cases. These cases, drawn from a range of sectors and organizational contexts, are designed to help readers envision real-world scenarios. They offer insights into potential opportunities and risks, suggest strategies for mitigation, and guide on evaluating outcomes. Through these examples, readers can better understand how to apply the principles discussed in this chapter to their own unique situations and challenges.

Sample Case 1: Strategic Generative Design Adoption

Industry: Industrial Manufacturing

Company Size: Large Conglomerate

Business Scenario: A heavy equipment manufacturer sought to harness generative design AI for rapid design iterations. However, isolated testing lacked scaled impact.

Solution: They instituted structured pilot programs across 5 business units with executive sponsorship, quantified efficiency KPIs and milestone-gated rollout plans. This ensured careful monitoring while aggressively scaling.

Outcomes and Impact: Broad deployment increased engineering productivity 43% over 18 months, enabling faster speed to market worth over $150 million revenue upside. Staged approach prevented quality issues.

Implementation Challenges: Customizing varied generative model integration with complex legacy design systems.

Sample Case 2: Phased Conversational AI Scaling

Industry: Telecommunications

Company Size: Mid-sized Organization

Business Scenario: A digital telco developed conversational AI to enhance customer support. However, company-wide rollout risks from the exploratory chatbot weren't assessed adequately.

Solution: They initially expanded the virtual assistant to 10% of subscribers, soliciting extensive user feedback revealing pain points before further expansion in phases.

Outcomes and Impact: The staggered approach smoothed adoption, preventing large-scale customer dissonance. By addressing revealed limitations in stages, efficiency lift reached projected 30% over two years.

Implementation Challenges: Quantifying scaling milestones based on fluid conversational AI metrics proved complex.

SAMPLE CASE 3: REGIONAL AI SOLUTION TRANSFER

Industry: Pharmaceuticals

Company Size: Large Multinational

Business Scenario: A global pharmaceutical enterprise successfully tested AI for accelerating molecular discovery in North America. However, directly adapting solutions organization-wide posed change management hurdles.

Solution: They instituted regional transition teams to reassess models against local datasets, tools and team skills ahead of staggered deployment using culturally resonant communication campaigns.

Outcomes and Impact: The regional tuning pre-empted clashes from imposed solutions. Teams could customize AI to their contexts resulting in over 40% efficiency gains realized globally in 18 months.

Implementation Challenges: Budget overruns from duplicated transition efforts. Streamlined through centralized governance.

Executive Sponsor: Role Specification

The Executive Sponsor is a senior leadership role responsible for overseeing the successful implementation and scaling of generative AI projects within the organization. This role requires a blend of strategic insight, leadership, and a deep understanding of the potential and challenges of generative AI technologies.

Key Responsibilities:

- Strategic Alignment: Ensure that the generative AI projects align with the organization's overall strategic objectives and business goals.

- Resource Allocation: Oversee and approve the allocation of necessary resources, including budget, personnel, and technology, for the AI projects.

- Stakeholder Management: Act as a primary liaison between the project team and other stakeholders, including the board, investors, and external partners.

- Risk Management: Monitor and address potential risks associated with the AI projects, ensuring that risk management strategies are in place.

- Project Advocacy: Champion the AI initiatives within the organization, advocating for support and understanding across different departments and teams.

- Performance Monitoring: Regularly review project progress against set goals and objectives, ensuring that the project delivers the expected value.

- Compliance and Ethics: Ensure that the AI projects comply with relevant laws, regulations, and ethical standards, particularly around data usage and privacy.

- Change Leadership: Lead change management efforts, addressing organizational resistance and fostering a culture that embraces innovation and AI-driven transformation.

Qualifications and Skills:

- Leadership Experience: Proven experience in a senior leadership role, preferably with exposure to technology or digital transformation initiatives.

- Understanding of AI Technologies: While not necessarily a technical expert, should have a solid understanding of AI and its business applications.

- Strategic Thinking: Ability to align technology initiatives with broader business strategies and objectives.

- Communication Skills: Strong communication and interpersonal skills to effectively manage stakeholder expectations and promote the AI projects.

- Risk Management: Experience in identifying and mitigating risks in technology projects.

- Change Management: Demonstrated ability in leading organizational change and innovation.

Personal Attributes:

- Visionary Leadership: Able to envision the transformative potential of AI and inspire others towards this vision.

- Adaptability: Comfortable with ambiguity and rapid changes typical in AI projects.

- Collaborative Approach: Ability to work collaboratively with various teams and stakeholders, fostering an environment of teamwork.

- Ethical Integrity: Commitment to upholding high ethical standards in all aspects of the AI initiative.

Reporting Structure:

The Executive Sponsor typically reports directly to the CEO or the board and works closely with the Chief Technology Officer, AI project managers, and department heads.

. . .

This role specification provides a guideline of what is required from an executive sponsor in generative AI projects. It can guide organizations in selecting the right candidate for this pivotal role, ensuring effective leadership and governance for their AI initiatives.

8

PERFORMANCE MONITORING AND CONTINUOUS IMPROVEMENT

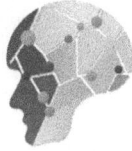

Harnessing generative AI's fluid capabilities for sustained impact requires vigilant tracking mechanisms and continuous improvement cycles. Unlike conventional software, generative AI exhibits probabilistic behaviors warranting close monitoring.

This chapter discusses vital strategies across two areas:

1. Quantifying generative AI value through relevant key performance indicators (KPIs)

2. Optimizing models through responsive feedback loops

Tracking Metrics

Metrics transform vague aspirations into executable generative AI success yardsticks. Aligning indicators to business priorities also elevates stakeholder trust.

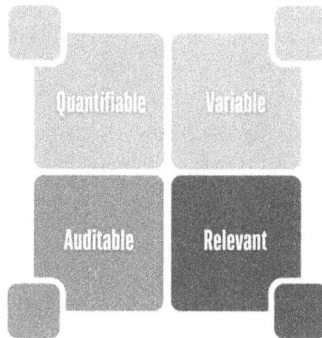

Effective KPIs exhibit properties like:

- Quantifiable: Rely on numerical data for tangibility aligned to motives like sales increases, risk reductions or product enhancement rates. Proxy values like usage intensity for intangibles supplement.

- Variable: Specify target outcomes, not just effort proxies like training durations. Outlining upside and downside variability also aids troubleshooting.

- Auditable: Support easy verification through linkage to trusted datasets rather than subjective human judgment vulnerable to manipulation.

- Relevant: Avoid vanity metrics like technical accuracy alone. Relevance ties directly to business benefits framed in end-user contexts like customer satisfaction or revenue impact.

Feedback Loops

Any generative AI application operates in dynamic environments. Users needs evolve, new datasets patterns emerge, regulations transpose—necessitating constant model updates.

Robust feedback pipelines enable continuous retraining:

- Proactive User Reviews: Instead of assuming stability, regularly eliciting qualitative user feedback highlights enhancements areas. Surfacing frustrations early, especially from frontline staff encountering systemic gaps behind company walls, is invaluable.

- Prediction Deviation Tracking: Probe quantitative divergences between generative model outputs and actual business metrics through statistical process control limits prompting updates before lagging effects worsen.

- Version Comparison Testing: Evaluate successive model variants thoroughly through techniques like A/B testing with control groups to certify improvements scientifically amid iterations. Prevents overcorrection risks.

Conclusion

Instilling generative AI environments with ingrained measurable value accountability and persistent tuning through multi-mode feedback flows sustains their functional potency while optimizing solution return on investment.

Sample Cases in Action

To bring the concepts of this chapter to life and demonstrate their practical application in various industries, I present a series of illustrative use cases. These cases, drawn from a range of sectors and organizational contexts, are designed to help readers envision real-world scenarios. They offer insights into potential opportunities and risks, suggest strategies for mitigation, and guide on evaluating outcomes. Through these examples, readers can better understand how to apply the principles discussed in this chapter to their own unique situations and challenges.

SAMPLE CASE 1: MONITORING SUPPLY FORECASTING AI

Industry: Online Retail

Company Size: Large Enterprise

Business Scenario: One of the world's biggest retailers developed AI-based supply and demand forecasting. However, model degradation risks from scale were inadequately tracked.

Solution: They instituted robust telemetry using deviation band alarming for key projections alongside continuous AB testing, allowing rapid detection and correction minimizing inventory impairment and stockout costs.

Outcomes and Impact: Over two years, the rigorous monitoring regimes alone yielded $75 million savings by nimbly adjusting for demand changes and preventing wastage.

Implementation Challenges: Clarifying accountability between central and regional analytics teams for monitoring.

SAMPLE CASE 2: GATHERING USER FEEDBACK ON AI RECOMMENDATIONS

Industry: Media Streaming

Company Size: Mid-sized Company

Business Scenario: A video streaming platform deployed neural collaborative filtering algorithms to personalize content recommendations. However, they faced difficulties assessing subjective aspects like relevance amid rapid content changes.

Solution: They implemented large-scale weekly customer surveys with precise questions around satisfaction with recommended shows while tracking streaming behavior correlation. This highlighted drop-offs needing model retraining.

Outcomes and Impact: The surveys and behavior tracking together optimized recommendations, lifting streaming consumption 32% and revenue $7 million over a year.

Implementation Challenges: Survey fatigue and falling participation over time. Mitigated via engagement incentives.

SAMPLE CASE 3: VERSION COMPARING TEXT GENERATION AI

Industry: Marketing Services

Company Size: Small Agency

Business Scenario: A boutique marketing firm tested cutting-edge AI copywriting software to boost content output. However determining true enhancement from experimental upgrades with limited data was challenging.

Solution: They manually A/B tested successive tool versions' output along dimensions like message resonance before migrating further, preventing regression.

Outcomes and Impact: The comparisons augmented quality assurance, allowing scaling written content 3X faster while preventing brand risks from unchecked AI created content.

Implementation Challenges: Quantifying subjective quality perceptions consistently across reviewers. Addressed through rater training.

Indicative KPIs and Measures for Generative AI Projects

Business Performance KPIs:

- ROI (Return on Investment): Measures the financial return on the AI project compared to its cost.

- Cost Savings: Reduction in operational costs due to AI implementation.

- Revenue Growth: Increase in revenue attributable to the AI project.

- Productivity Improvements: Gains in efficiency and output due to AI adoption.

Customer Impact Measures:

- Customer Satisfaction Scores: Changes in customer satisfaction levels post-AI implementation.

- Net Promoter Score (NPS): Measures customer loyalty and likelihood of recommending your product/service.

- Customer Engagement Metrics: Engagement levels on digital platforms, including time spent, interaction rates, etc.

- Customer Retention Rate: The impact of AI initiatives on retaining existing customers.

People (Employees and Society):

- Employee Satisfaction and Engagement: Employee attitudes towards the AI initiatives and changes in their work environment.

- Training and Development: Investment in employee training for AI-related skills and new roles.

- Diversity and Inclusion Metrics: Impact of AI on workforce diversity and inclusivity.

- Social Impact: Assessment of how the AI project affects the wider community and society.

Planet (Environmental Impact):

- Carbon Footprint: The environmental impact of AI initiatives, particularly in terms of energy consumption and carbon emissions.

- Sustainable AI Practices: Adoption of eco-friendly practices in AI development and deployment.

- Waste Reduction: Impact of AI on reducing waste, both in digital and physical processes.

Partners (Suppliers and External Stakeholders):

- Partner Satisfaction: Feedback and satisfaction levels from suppliers and external partners involved in the AI project.

- Collaboration Effectiveness: Effectiveness of collaborations and joint initiatives with partners.

- Supply Chain Efficiency: Improvements in supply chain processes due to AI integration.

- Ethical Supply Chain Compliance: Adherence to ethical practices in the supply chain influenced by AI.

General AI-Specific Measures:

- Model Accuracy and Performance: The effectiveness of AI models in real-world applications.

- AI Ethics Compliance: Adherence to ethical guidelines in AI development and deployment.

- Data Privacy and Security Metrics: Effectiveness of data protection and privacy measures in AI systems.

- Innovation Rate: The rate of new solutions or improvements generated by AI initiatives.

These KPIs and measures provide a comprehensive framework for organizations to evaluate the success and impact of their generative AI projects from multiple dimensions. By covering business, customer, people, planet, and partner aspects, organizations can gain a holistic understanding of their AI initiatives, ensuring that they are not only profitable but also socially responsible and sustainable.

Understanding Large Language Model Benchmarks

In the swiftly evolving landscape of generative AI and Large Language Models (LLMs), benchmarks have become indispensable tools. They provide essential insights into the capabilities and limitations of these advanced technologies. Here, I cover the importance of benchmarks in the realm of AI, with a particular focus on the contributions from Stanford's Human-Centered Artificial Intelligence (HAI) and Hugging Face. I explore their relevance and implications for both business leaders and AI developers, offering a deeper understanding of these critical evaluative tools.

The Role of Benchmarking in AI

Benchmarks in AI serve as standardized metrics to evaluate the performance of AI models. These benchmarks are not just about scoring or ranking; they provide a comprehensive view of how models perform under different scenarios and tasks. By offering a standardized way to measure and compare, benchmarks are invaluable for tracking progress and guiding future developments in AI technologies.

Major Benchmarks in Generative AI

Recent years have seen the development of significant benchmarks in the field of generative AI. Notably, Stanford HAI's Center for Research on Foundation Models introduced the Holistic Evaluation of Language Models (HELM). This benchmark aims to improve the transparency and understanding of language models. Simultaneously, Hugging Face's Open LLM Leaderboard tracks and evaluates open LLMs and chatbots. This leaderboard is distinguished by its comprehensive approach, incorporating various components such as Chatbot Arena, MT-Bench, and MMLU (5-shot), each focusing on different aspects of language model performance.

Implications for Business and Development

For business leaders and AI developers, these benchmarks are more than just evaluative tools; they are guides for strategic decision-making and technical enhancement. Understanding these benchmarks helps business leaders make informed decisions about AI adoption and investment. For developers, they provide a roadmap for improving model accuracy, efficiency, and scalability.

Challenges in Benchmarking

While benchmarks provide valuable insights, they also come with limitations. One of the primary challenges is ensuring that these benchmarks accurately represent the diverse scenarios in which AI models operate. Additionally, as AI technology continues to evolve, benchmarks must adapt to assess new capabilities and address emerging ethical concerns.

The Future of AI Benchmarking

Looking ahead, the field of AI benchmarking is set to evolve significantly. Innovations in AI technology will require the development of new benchmarks, capable of evaluating more complex and nuanced aspects of AI models. This evolution will be crucial for maintaining the relevance and effectiveness of benchmarking in assessing AI technologies.

Conclusion

Benchmarks play a crucial role in the field of generative AI and LLMs, offering insights that guide both development and strategic application. As AI technologies continue to advance, understanding these benchmarks will be pivotal for anyone involved in the AI landscape, be it in a technical or managerial capacity.

9

ORGANIZATIONAL CULTURE AND CHANGE MANAGEMENT

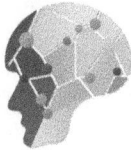

Realizing generative AI's transformative potential ultimately hinges on people. Without earnest efforts focused on cultural readiness and change management, inertia can readily undermine technological promise.

This chapter discusses two pivotal elements for activation:

1. Securing leadership commitment to sponsor generative AI programs

2. Preparing the workforce to adopt new AI-powered processes

Leadership Involvement

Attempting enterprise-wide generative AI progress sans executive engagement courts failure through budgetary apathy or misaligned priorities across middle management stuck in status quo.

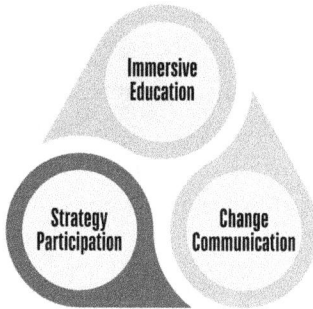

Getting leadership effectively on board involves:

- Immersive Education: Beyond cursory presentations, facilitation through workshops, external site visits and hands-on experimentation crystallizes AI implications allowing intuitive advocacy.

- Strategy Participation: Directly involving executives in charting AI vision and governance via advisory councils or hackathons rather than post-hoc updates seeds authentic engagement, quickening supportive policymaking.

- Change Communication: Leaders publically embracing AI transformation through townhalls, interview messaging and social posts provides cultural signaling instrumental for driving adoption momentum bottom-up.

Change Management

As workflows and decision dynamics transform with generative AI infusion, failure to support people struggling with job uncertainty fears or capability gaps manifests in resentment-fueled implementation resistance.

Thriving adoption requires:

- Workflow Impact Analysis: Clarity on how specific roles get redefined or reconfigured through process reengineering aids customized transition plans including re-skilling.

- Empathetic Communication: Transparency through empathetic dialog instead of evasive assurances or denials catalyzes trust, surfacing adoption barriers needing redressal.

- New Capability Support: From AI User Experience labs to chatbot assistants, surround workforce with sustained guidance easing generative AI onboarding despite learning curve strains.

Conclusion

With cultural readiness cemented through leadership motivation and workforce change management, generative AI flux gets transformed into an accelerative competitive advantage owned by empowered teams.

Sample Cases in Action

To bring the concepts of this chapter to life and demonstrate their practical application in various industries, I present a series of illustrative use cases. These cases, drawn from a range of sectors and organizational contexts, are designed to help readers envision real-world scenarios. They offer insights into potential opportunities and risks, suggest strategies for mitigation, and guide on evaluating outcomes. Through these examples, readers can better understand how to apply the principles discussed in this chapter to their own unique situations and challenges.

SAMPLE CASE 1: EXECUTIVE AI IMMERSION WORKSHOP

Industry: Automotive

Company Size: Large Manufacturer

Business Scenario: The CEO of an auto major sought deeper generative tech understanding to direct strategic roadmaps. However securing time for immersion was challenging amid business priorities.

Solution: A dedicated 2-day offsite workshop got designed leveraging external site visits, tailored presentations, hands-on labs and expert panel discussions yielding multi-faceted insights on possibilities, guardrails and success factors shaping conviction.

Outcomes and Impact: Post-workshop, the executive greenlit a $50 million investment into autonomous driving research applications, catalyzing long-term growth prospects.

Implementation Challenges: Preventing discussions from getting distracted by short-term business metrics.

SAMPLE CASE 2: WORKFORCE ENABLEMENT FOR CONVERSATIONAL AI

Industry: Healthcare

Company Size: Mid-sized Hospital

Business Scenario: A specialty hospital planning patient service AI rollouts faced adoption risks from staff wary of job impacts from automation. However leadership backed the solutions as competitiveness imperatives.

Solution: They proactively conducted empathetic listening circles with affected teams, published clear role evolution guides and instituted on-demand microlearning support channels easing the transition.

Outcomes and Impact: The supportive interventions smoothed staff acceptance, allowing cutting AI operational-ization time nearly 40% while preventing talent drain.

Implementation Challenges: Quantifying productivity return on change management investments. Optimized via OKR techniques.

SAMPLE CASE 3: GENERATIVE AI STARTUP SUCCESSION PLANNING

Industry: Venture Capital

Company Size: AI Startup

Business Scenario: The founding engineer of an AI platform startup wanted to scale business operations. However key generative IP and model know-how remained concentrated risking investor confidence.

Solution: They deliberately cross-trained junior team members through knowledge transfers, built process play-books capturing institutional wisdom and prepared leader-ship backups through external coaching.

Outcomes and Impact: The interventions boosted investor confidence on business resilience realizing $15 million Series B fundraising within 10 months to accelerate product boosting equity value 3X.

Implementation Challenges: Motivating expert teams to conduct knowledge transfers diluting uniqueness. Improved through vision alignment.

Research Insights: Automation and Workforce Impacts

In the transformative journey of generative AI, understanding its impact on the workforce is critical. This section synthesizes insights from previous research and recent studies, providing a nuanced view of the evolving work landscape. These should be considered in change plans.

1. Automation Blindness and Human Oversight[1]

The integration of AI systems into the workforce has introduced the risk of 'automation blindness,' where human workers become passive monitors rather than active participants. This phenomenon can lead to oversight failures and errors in judgment. To counteract this, it is essential to engage with AI's internal reasoning processes, visualize confidence levels, and regularly prompt human acknowledgment of AI actions. These interventions can help maintain an active role for human workers, fostering a balanced relationship with AI systems.

2. Costs of Overly Capable Automation[2]

Highly accurate AI systems can paradoxically reduce human attentiveness and diligence, especially among seasoned professionals. To maintain human engagement, it may be beneficial to use less accurate AI, thereby encouraging workers to compensate for AI shortcomings. Additionally, tailoring the level of automation to the skill levels of users and incentivizing human discretion can ensure that human judgment remains a vital part of the decision-making process.

3. Impact on Teamwork and Social Dynamics[3]

The introduction of AI into team environments can disrupt established collaborative dynamics. While AI optimizes individual tasks, it may overlook the nuances of group coordination and social interaction. To successfully integrate AI into teams, a holistic approach is necessary. This includes custom training programs and AI systems designed to support, rather than hinder, human collaboration.

4. Social Loafing in Redundant Monitoring[4]

Redundant monitoring systems, designed to enhance reliability through teamwork, can inadvertently lead to social loafing, where individuals in a team exert less effort. To counteract this, strategies such as maintaining individual accountability and creating targeted incentives can be effective. These approaches encourage each team member to contribute actively and meaningfully.

5. Generative AI and Workforce Impact

Recent research, including studies by McKinsey[5] and the International Labour Organization (ILO)[6], highlights that generative AI is set to automate up to 30% of work hours in the US by 2030. While it's expected to augment rather than eliminate jobs in fields like STEM and creative professions, significant impacts are anticipated in office support, customer service, and food service sectors. Women, particularly in clerical roles, are likely to be disproportionately affected due to the gendered nature of these jobs. Workers in lower-wage positions face the prospect of significant occupational shifts and will require re-skilling to transition successfully. Generative AI's potential to boost labor productivity also presents an opportunity for economic growth, provided that worker transitions and associated risks are effectively managed.

. . .

Successful integration of automation and AI into the workforce requires a conscious and holistic approach. Key strategies include promoting human engagement and attentiveness, customizing AI capabilities to user needs and roles, engineering systems for teamwork enhancement, motivating discretion and judgment over blind reliance on AI, and closely monitoring impacts on social dynamics and teamwork.

These studies outline the intricate balance required in the evolving workplace, where the capabilities of generative AI and automation must be harnessed without undermining the unique strengths and contributions of human workers. They emphasize the need for proactive policies and practices that support workforce adaptation to these technological advancements, ensuring that the benefits of higher productivity and innovation are broadly shared.

---

1. Harvard Business Review: The Tragic Crash of Flight AF447 Shows the Unlikely but Catastrophic Consequences of Automation by Nick Oliver, Thomas Calvard, and Kristina Potočnik.
2. Falling Asleep at the Wheel: Human/AI Collaboration in a Field Experiment on HR Recruiters by Fabrizio Dell'Acqua.
3. Super Mario Meets AI: Experimental Effects of Automation and Skills on Team Performance and Coordination by Fabrizio Dell'Acqua, Bruce Kogut, and Patryk Perkowski.
4. Human Redundancy in Automation Monitoring: Effects of Social Loafing and Social Compensation by Juliane Domeinski, Ruth Wagner, et al.
5. "Generative AI: How will it affect future jobs and workflows?" by McKinsey Global Institute.
   "Generative AI and the future of work in America." by McKinsey Global Institute.
6. "Generative AI and jobs: A global analysis of potential effects on job quantity and quality." by Paweł Gmyrek, Janine Berg, David Bescond.

10

FUTURE-PROOFING

Generative AI represents an intensely dynamic domain, with new techniques and paradigms emerging continuously. In such an environment, relying solely on current capabilities risks rapid obsolescence as innovations rewrite best practices.

This chapter discusses two crucial strategies for sustained relevance:

1. Committing resources expressly for generative AI research and development

2. Embedding organizational agility to adapt to AI progress

Innovation and R&D

Maintaining an innovation pipeline through R&D future-proofs organizations against disruption from new generative AI advances.

Elements for viability include:

- Structured Programs: Dedicate budgets, data and talent expressly towards blue-sky programs exploring high-risk, high-reward ideas beyond short-term implementation horizons through autonomous research units.

- University Partnerships: Collaborations spanning sponsored academic research to staff exchange programs, hackathons and published papers provide intellectual exchange vital for pushing boundaries.

- Start-up Engagements: Incubating early-stage generative AI startups through investments, accelerators and commercial pilots seeds exposure to emerging techniques with mutual innovation opportunities.

Agile Adaptation

Equally crucial as inventing frontier concepts is operational readiness to rapidly assimilate cutting-edge generative AI capabilities through deliberate flexibility.

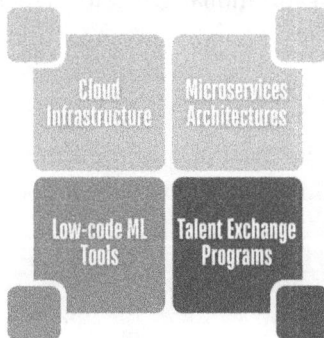

Enablers encompass initiatives like:

- Cloud Infrastructure: Hardware abstraction through cloud platforms allows seamless scale-up for new model demands unlike on-premise data center overhauls.

- Microservices Architectures: Componentized services ease integration of improved AI modules or third-party algorithms into existing application stacks.

- Low-code ML Tools: Empowering business teams to directly build models, tweak data and iterate relieves change burdens on central AI units, aiding experimentation.

- Talent Exchange Programs: Facilitating expert staff rotations across strategic university/startup partnerships diffuses emerging skill sets.

Conclusion

In dynamic environments, it is insulated organizations which get disrupted the most. By consciously fostering pioneering R&D and change-adept cultures, companies harness generative AI with resilient, future-proof competitive advantage.

Sample Cases in Action

To bring the concepts of this chapter to life and demonstrate their practical application in various industries, I present a series of illustrative use cases. These cases, drawn from a range of sectors and organizational contexts, are designed to help readers envision real-world scenarios. They offer insights into potential opportunities and risks, suggest strategies for mitigation, and guide on evaluating outcomes. Through these examples, readers can better understand how to apply the principles discussed in this chapter to their own unique situations and challenges.

SAMPLE CASE 1: FOCUSED GENERATIVE AI RESEARCH LAB

Industry: Semiconductors

Company Size: Large Company

Business Scenario: A semiconductor leader facing risks from slowing Moore's Law wanted to spur manufacturing innovations through generative AI and quantum techniques. However, short-term business horizons hindered blue-sky pursuits.

Solution: They launched an autonomous Advanced Research Lab with ring-fenced budget, talent and leadership oversight exploring cutting-edge areas like materials informatics, quantum assisted AI and robotic automation promising disruptive advances within 5-10 year horizons.

Outcomes and Impact: The lab has produced over 24 patents around pioneering applications, establishing innovation leadership and preempting competitive risks.

Implementation Challenges: Quantifying returns from abstract innovation pursuits. Optimizing through external advisory reviews.

Sample Case 2: Rapid Prototyping of Third-Party AI Innovations

Industry: Business Services

Company Size: Mid-sized Company

Business Scenario: A management consulting firm actively tracked startups with exciting generative AI solutions relevant to their services. However lengthy evaluations stalled integration.

Solution: They instituted specialized Microventures Teams to rapidly prototype external innovations through focused data sharing and app development sprints crystallizing capability enhancement potential ahead of strategic investments or partnerships.

Outcomes and Impact: The approach has halved due diligence cycles while derisking evaluations. So far, two startups have been acquired integrating cutting-edge expert recommendation engines into solution portfolio.

Implementation Challenges: Overcoming internal cultural inertia towards external innovations. Improved through executive directives.

Sample Case 3: Proactive Re-Training for AI Model Updates

Industry: Financial Services

Company Size: Mid-sized Firm

Business Scenario: A digital lending fintech startup periodically updated its credit risk prediction model to prevent data drift. However, conducting full regression testing repeatedly bogged down product teams delaying feature enhancement.

Solution: They implemented automated unit testing, capabilities-based staff rotation and provisioned ramp-up buffers allowing rapid assimilation of updated models through Delaware releases without major business disruption.

Outcomes and Impact: The future-proofing initiatives compressed regression testing cycles by over 60%, catalyzing faster iteration of new product offerings.

Implementation Challenges: Quantifying trade-offs between scaled innovation pace facilitated versus opportunity costs of excess capacity buffers.

Framework for Building Exploratory Innovation Capability

Here is a framework for building exploratory innovation capability, particularly in generative AI:

- **Establish Innovation Labs:** Create specialized labs or units dedicated to pure research in generative AI. These should be environments where exploratory thinking is encouraged, and traditional business pressures are minimized.

- **Promote Interdisciplinary Teams:** Assemble teams from various disciplines such as AI, data science, psychology, and design. The diversity in backgrounds can spur creative solutions and innovative approaches.

- **Engage in Open Innovation:** Utilize platforms for crowdsourcing ideas and collaborate with external entities like universities, startups, and research institutes. This can expand the pool of ideas and introduce novel perspectives.

- **Conduct Idea Generation Workshops:** Regularly organize brainstorming sessions focused on blue-sky thinking. Encourage participants to think beyond existing constraints and envision future possibilities with generative AI.

- **Implement Flexible Resource Management:** Develop a system for dynamically allocating resources to exploratory projects. This allows for quick pivoting and scaling up of promising initiatives.

- **Redefine Success Metrics for R&D:** Establish new metrics that focus on long-term innovation and knowledge creation, such as the number of patents filed, research papers published, or new methodologies developed.

- **Foster a Culture of Experimentation:** Cultivate an organizational ethos that values risk-taking, accepts failure as a learning opportunity, and celebrates creative problem-solving.

- **Continuous Learning and Skill Development:** Encourage ongoing education and training in emerging technologies and methodologies to keep the R&D team at the forefront of generative AI developments.

- **Regular Review and Adaptation:** Continuously assess the effectiveness of the R&D activities and adapt strategies based on feedback and emerging trends in the field of generative AI.

These suggestions aim to help you build a robust, dynamic R&D capability in generative AI, balancing creative exploration with strategic alignment to the organization's long-term goals.

11

THE WAY FORWARD

As we reach the concluding pages of this blueprint, let's reflect on the generative AI transformation journey so far.

From foundational building blocks of awareness, strategy and skills to change management bridges empowering adoption, each chapter spotlighted key enablers for harnessing generative AI responsibly to drive business growth.

Synthesizing insights from advisory leaders and research, the evidence-based guidance also cautions against hype-swayed decisions, emphasizing balanced perspectives attuning investments to genuine use case viability.

Nonetheless, amid all complex technology and cultural interventions, the single most crucial factor underpinning successful generative AI integration remains leadership resolve to firmly yet adaptively commit to the transformation journey without wavering at inevitable ambiguity waypoints.

Much as the first industrial revolution held challenges alongside promise, so too will this new era led by exponential technologies like AI. With eyes wide open to risks but also possibilities, the strategies here equip teams to traverse the challenges ultimately to tap substantial upsides.

As AI pioneer Andrew Ng famously quipped, AI is the new electricity. Companies unable to harness its power risk disappointing customers seeking AI-enabled convenience while competitors outpace them. However, organizations which view implementation hurdles as surmountable challenges can claim decisive competitive advantage.

Parting Thoughts

Generative AI is a technological inflexion point promising immense opportunity. Realizing its potential warrants undertaking an exploratory quest filled with experimentation, discovery and meaningfully addressing risks like reliability gaps or job losses which understandably seed skepticism today just as automation or any new technology spawned initially.

With pragmatic frameworking, the turbulence settles gradually into ecosystems where AI and people collaborate symbiotically to elevate human potential. That ultimately remains the north star guiding this transformation.

When adopted holistically and responsibly across the key areas detailed here—people, process, technology, governance, and planet—generative AI can securely unlock unprecedented progress in coming years. The blueprint now lies mapped.

Bye for now, and stay human!

Inês

END

Note: Consider leaving a review and sharing the book with others that may benefit from its content.

I encourage you to dive deeper into the technology powering generative AI by reviewing the books and courses on the next page.

KEEP LEARNING

Our books

Our courses